Michael Bestehorn

Non-Equilibrium Thermodynamics

Also of Interest

Thermal Analysis and Thermodynamics
In Materials Science
Detlef Klimm, 2022
ISBN 978-3-11-074377-7, e-ISBN (PDF) 978-3-11-074378-4

Differential Equations, Fourier Series, and Hilbert Spaces
Lecture Notes at the University of Siena
Raffaele Chiappinelli, 2022
ISBN: 978-3-11-129485-8; e-ISBN 978-3-11-130252-2

Non-equilibrium Thermodynamics and Physical Kinetics
Halid Bikkin and Igor I. Lyapilin, 2021
ISBN 978-3-11-072706-7, e-ISBN (PDF) 978-3-11-072719-7

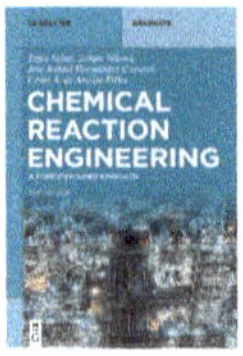

Chemical Reaction Engineering
A Computer-Aided Approach
Tapio Salmi, Johan Wärnå, José Rafael Hernández Carucci and César A. de Araújo Filho, 2023
ISBN 978-3-11-079797-8, e-ISBN (PDF) 978-3-11-079798-5

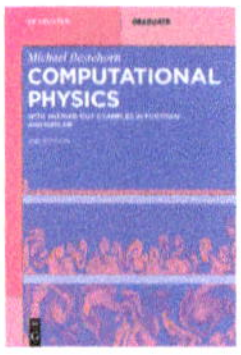

Computational Physics
With Worked Out Examples in FORTRAN® and MATLAB®
Michael Bestehorn, 2023
ISBN 978-3-11-078236-3, e-ISBN (PDF) 978-3-11-078252-3

Michael Bestehorn

Non-Equilibrium Thermodynamics

Entropy, Heat and Mass Transfer

DE GRUYTER

Author
Prof. Michael Bestehorn
Brandenburg University of Technology
Cottbus-Senftenberg
Dep. of Statistical Physics and Nonlinear Dynamics
Erich-Weinert-Str. 1
03046 Cottbus
Germany
bestehorn@b-tu.de

ISBN 978-3-11-079404-5
e-ISBN (PDF) 978-3-11-079413-7
e-ISBN (EPUB) 978-3-11-079419-9

Library of Congress Control Number: 2024931816

Bibliographic information published by the Deutsche Nationalbibliothek
The Deutsche Nationalbibliothek lists this publication in the Deutsche Nationalbibliografie; detailed bibliographic data are available on the Internet at http://dnb.dnb.de.

Cover image: Michael Bestehorn
Typesetting: VTeX UAB, Lithuania
Printing and binding: CPI books GmbH, Leck

www.degruyter.com

I could have loved you, girl, like a planet
I could have chained your heart to a star
But it really doesn't matter at all
No it really doesn't matter at all
Life's a gas

Marc Bolan 1971

Preface

The development of thermodynamics, or more precisely "phenomenological thermodynamics", began in the early 19th century with the invention and extensive operation of steam engines. Contrary to classical point mechanics, devised by Isaac Newton and his followers starting in the 17th century, thermodynamics is based on the assumption that matter, like a fluid, a gas, or a solid, has the form of a deformable continuum that fills the whole space. Such a continuum can normally be described by very few material parameters (heat capacity, density, elasticity, etc.), and as a consequence its present state is defined by very few state variables, such as temperature or pressure. The description of processes where heat and its transformation to mechanical work play a crucial role was not feasible in the frame of Newton's mechanics. Later it turned out that the concepts of thermodynamics are very general and can be applied to other scientific branches, such as chemical reactions, biological, or nowadays even social questions.

Students of sciences or engineering normally study a basic course of one or two semesters on thermodynamics, restricted to "processes" in thermal equilibrium. In fact, the quotation marks denote a contradiction: time plays essentially no role in equilibrium physics, because one of the definitions of equilibrium is time independence of all variables. But "process" means variation and transformation from a given state to another one, and processes in thermal equilibrium can at most "run" infinitely slowly. At least for the normally taught equilibrium thermodynamics, the notion "thermostatics" would be much more appropriate.

On the other hand, all natural processes, biological, astrophysical, or geophysical, to name but a few, take place far from equilibrium. Life itself is a constant fight against equilibrium that is eventually, but unavoidably, reached after death. Thus studying non-equilibrium thermodynamics seems not only be important for the physicist.

Processes far from equilibrium normally run in a unique direction; they are called "irreversible." A precious wine glass falls down from the table and breaks into pieces while its potential energy is finally transformed in heat energy, heating up the pieces and the surrounding air. The inverse process: the air and the pieces cool down and put their heat energy first into the creation of the very same antique glass and later into kinetic energy needed to lift the glass from the floor back to the table, can be excluded. Both processes are possible from the point of energy conservation, and both are far from equilibrium. But the broken pieces on the floor are much closer to equilibrium than the glass on the table. In the first case, equilibrium is approached; in the reverse one it would be abandoned.

A measure for the distance from equilibrium is entropy. The closer a system is to equilibrium, the higher is its entropy. A process running versus equilibrium will thus act as a source for entropy. It produces entropy as long as it is out of equilibrium. We shall call such a process a "natural process." Of course the backwards process to a natural process can be performed, but a lot of machinery and manpower, for our example, amongst

https://doi.org/10.1515/9783110794137-201

others, an experienced glassblower, would be necessary. At the end, this endeavor would produce much more entropy as that lost from reconstruction of the glass on the table.

But how is entropy defined? First introduced by Rudolf Clausius in 1865, entropy is a state function as energy and at least its changes can be measured and computed. Every thermodynamic system has a certain value of entropy, which is unique for a given state. In a closed (isolated) system, all natural processes increase the total entropy. The entropy reaches a maximum if the system is in equilibrium. This is the second law of thermodynamics. Entropy and its production is a central concept in non-equilibrium thermodynamics. The distance from equilibrium is proportional to the amount of entropy production per time. But this distance can also be interpreted as the source for generalized forces that create fluxes, which drive the system towards equilibrium. The fluxes are associated with densities, such as energy and momentum of the continuum by conservation laws. The conservation laws have the form of partial differential equations and are well known, for instance, as heat equation, Navier–Stokes equations, and continuum equation, or in multi-component systems, as reaction-diffusion equations.

Thermodynamics can be considered as the science of macroscopic objects, where the microscopic or atomic level plays no role. However, if the microscopic world would be ruled by classical mechanics, Newton's equations of motions should apply for the microscopic particles. Such a hypothesis indicates a bridge between Newton's microscopic world and thermodynamic theories. This bridge was identified in the second half of the 19th century, mainly by James Clerk Maxwell and Ludwig Boltzmann. Heat was defined as internal energy of microscopic particles and temperature as the mean deviation of particle velocities from a mean velocity. Boltzmann connected the disorder of a microscopic gas to the value of the so-called H-functional that plays the role of the macroscopic (negative) entropy. With his nowadays called "H-theorem," he could prove that the H-functional in a closed system always decreases, putting the second law on purely mechanical grounds. In this way, Boltzmann could solve the old riddle how on the microscopic level a bunch of gas particles moving reversibly can lead to irreversible macroscopic behavior with a pronounced direction of the time arrow. Here we only mention a serious objection against Boltzmann's reasoning raised by Josef Loschmidt in 1876: What if one considers a gas of many particles on its way to equilibrium and suddenly *reverse all* particle velocities? If the microscopic physics is reversible, the gas then must move "backwards" towards states with lower and lower entropy and depart from equilibrium. This apparent contradiction to the H-theorem (and to the second law) is nowadays known as "Loschmidt's paradox," "irreversibility paradox" or "Umkehreinwand." For details and its "solution" we refer to the discussion in Chapter 8.

Energy and entropy are the two pillars of all natural sciences, and hence far from being only important in thermodynamics. Each is ruled by one of the two fundamental thermodynamic laws. If we assume that our world is closed and isolated, the two laws have very general consequences:

(I) The total energy of the universe is constant.
(II) The total entropy of the universe can never decrease.

In the long time limit, the universe will come to thermal equilibrium and all energy will be transformed into heat in form of radiation. The formation and the existence of patterns such as those of clouds, waves, stones, plants or animals can only be temporary, accounting for a locally limited decrease of entropy. Nevertheless, the total entropy thereby increases.

From the brief sketch outlined above several important notions and keywords can be gathered. The present textbook tries to explain these fields and put them into a common frame. A sort of assignment of the keywords to the book chapters is the following:

- state of a system and state variables, Chapter 1;
- internal energy and heat, Chapters 1–4;
- thermal equilibrium and non-equilibrium, Chapters 1, 2;
- irreversible processes, Chapters 2–6;
- entropy and irreversibility, Chapters 1, 2;
- conservation laws, Chapters 2, 3;
- self-organized pattern formation, Chapters 5, 6;
- bridge between microscopic (Newton's) and macroscopic world, Chapters 7–9.

The reader should be acquainted with the subject of equilibrium thermodynamics, as it is taught for instance in a basic course of theoretical physics in a bachelor degree programme. Many examples are studied and applications from fluid dynamics and chemical non-equilibrium reactions are discussed in detail. In the second part, the phenomenologically derived transport equations from the first part are put on statistical grounds following Boltzmann's kinetic theory of gases.

The rest of the book is organized as follows: Chapter 1 repeats the prerequisites, fundamental laws, and notions from equilibrium thermodynamics. Chapter 2 is the central chapter of the first part of the book and introduces conservation laws and a formula for entropy production, valid outside of equilibrium. Multi-component systems, i. e., mixtures of gases or fluids and chemical reactions, are thereby discussed in detail. in Chapter 3, the famous Onsager relations are applied in relation to conservation equations, and a set of partial differential equations or transport equations is presented. Chapters 4 to 6 show many applications coming from heat conduction, diffusion processes, chemical reactions, and hydrodynamics. Thereby, the emergence of instabilities followed by self-organized macroscopic pattern formation occurs quite naturally.

The second part is based on the kinetic gas theory devised by Ludwig Boltzmann and gives a statistical foundation for the equations and relations illustrated in the first part of the book. Starting with some general considerations of a classical many particle system in Chapter 7, the Boltzmann equation is stated and the H-theorem is proven in Chapter 8. In Chapter 9, the transport equations phenomenologically outlined in Chapter 3 are derived applying the relaxation time approximation. Finally, numerical solutions of the Boltzmann equation are presented for two simple problems and compared with solutions of the transport equations of the first part.

Contents

Part I: **Phenomenological thermodynamics**

Rudolf Julius Emanuel Clausius (1822–1888). Founder of the concept of entropy.
Picture from about 1850, unknown author, source: Wikimedia.

https://doi.org/10.1515/9783110794137-001

1 Equilibrium thermodynamics: a brief summary

We give a brief overview on the main ideas, concepts, and notions of the subject of equilibrium thermodynamics. We recapitulate the basic laws of thermodynamics and show their relevance for simple systems, such as perfect gases or perfect mixtures. Thereby, special emphasis is layed on irreversible processes and on the concept of entropy, a thermodynamic quantity primary to non-equilibrium thermodynamics. For readers not so familiar with the basics we refer to textbooks, which are many, e. g., [1, 2], to name only two.

1.1 State, variables, and thermal equilibrium

1.1.1 Thermodynamic system

In the context of thermodynamics, a system in its most simple form consists of a certain number of particles (or moles) of a gas, separated or at least distinguished somehow from its environment. But the definition of a system can be even more abstract. Its limitation is not necessarily a physical one, such as a wall or a membrane, it could also be a conceptual one, for instance an ocean, the atmosphere, or our solar system. However, in thermodynamics a system consists normally of a gas, a liquid, and/or a solid separated clearly from its environment.

1.1.2 State of a system and state variables

The state of a system is characterized by a set of variables, called state variables. Knowing the values of all state variables, the state of the system is completely known. The more complex a system is, the more such state variables are needed. Functions of one or more state variables are called state functions. For a certain given state, state variables and state functions take unique values. The state variables and the state functions can be grouped in

- *intensive variables*, such as temperature (T), density (ρ), pressure (p), conductivities, electromagnetic fields;
- *extensive variables*, such as energy (U, E), volume (V), mass (M), mole numbers (N_k, n_k), entropy (S).

Intensive variables do not depend on the system size. Extensive variables are normally written with capital letters. If they refer to one mole (or any other mass unit) or to a volume unit, small letters are used. Example:

$$s = \frac{dS}{dV}, \quad \hat{s} = \frac{dS}{dM} = \frac{dS}{dV}\frac{dV}{dM} = \frac{s}{\rho}. \tag{1.1}$$

https://doi.org/10.1515/9783110794137-002

Here, s is the entropy density (entropy related to volume) and $\hat{s}$ is the specific entropy (related to mass). Both s and $\hat{s}$ are intensive variables.

A *simple system* (one component) is completely described by only two independent state variables. A gas with a constant number of moles (particles) is a simple system.

1.1.3 Equations of state

Only a certain number of state variables can be chosen independently; this number is called number of degrees of freedom and can be computed by the phase rule of Gibbs:

$$F = C - P + 2 .$$

Here, C denotes the number of components and P the number of phases. For a one-component gas, $C = 1$, $P = 1$, and $F = 2$; for a binary gas mixture $C = 2$ and $F = 3$. The other state variables are linked by the equations of state (EOS). The EOS can be derived (or motivated) from experiments phenomenologically, by numerical models or by statistical physics. They have the general nonlinear form

$$F_i(p, V, T, N_k, \dots) = 0, \quad U = U(p, V, T, N_k, \dots) .$$

The first set is called thermal EOS; the latter one is the caloric EOS.

Example: The perfect gas equation

$$NRT - pV = 0$$

is a thermal EOS (R = gas constant, N = mole number). The caloric EOS for the perfect gas reads

$$U = Nc_V T$$

with the specific heat c_V. Here U is a state function, the internal energy.

Depending on the system under consideration, there can be more than one thermal state equation.

1.1.4 Processes

For a running process, the state variables change their values in time. It is important to determine processes that run by themselves (water flows down a river, a gas expands if it is exposed to a larger bin, two gases mix if they share the same volume), and those where a certain machinery is needed. Processes running on their own go always in a certain direction. Reversal of this direction would contradict to all previous observations. These processes are called *irreversible*. An irreversible process running on its own will stop when thermal equilibrium is reached. The distance from thermal equilibrium can be seen as a kind of driving mechanism for irreversible processes.

1.1.5 Thermal equilibrium

A closed system (mechanically and thermally isolated) reaches in the long time limit thermal equilibrium. Equilibrium is achieved by irreversible processes inside the system. In equilibrium, all state variables are constant in time. They can be
- *homogeneous in space*, in which case, they have everywhere the same value;
- *space dependent*; this is called inhibited equilibrium;

see Figure 1.1. In equilibrium, only reversible processes are possible. As long as the system is not in equilibrium, the state variables may be functions of time and space. They are described by partial differential equations, which follow in the optimal case from basic conservation laws (energy, momentum, mass), and/or from model assumptions (chemical reactions). Deriving these equations from first principles using the concept of thermodynamics is the main subject of non-equilibrium thermodynamics.

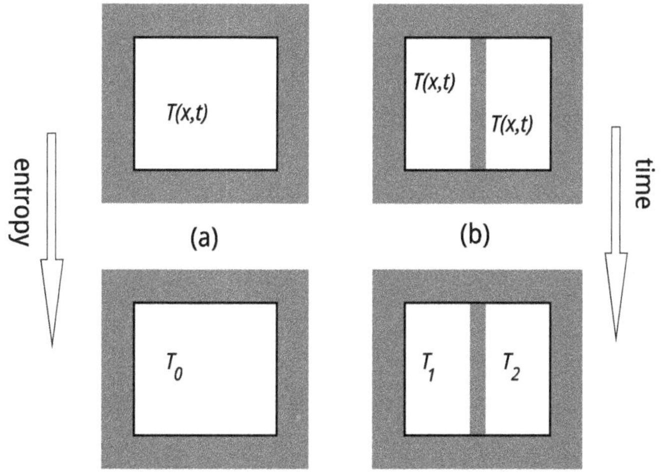

Figure 1.1: An isolated (closed) system on its way to thermal equilibrium: (a) the initial temperature depends on space and reaches at the end a uniform value due to irreversible processes; (b) the two inner systems are not in thermal contact and may end with different, but constant temperatures due to different internal processes. The final situation is named inhibited equilibrium.

Equilibrium thermodynamics can only deal with reversible processes, where the system is always in equilibrium. This is an idealization, because normally all processes in nature are irreversible. Reversible processes take an infinitely long time!

State variables and state functions are only defined in thermal equilibrium.

1.2 Energy and the first law

1.2.1 Internal energy

The (internal) energy U of a system is a state function. For every state determined uniquely by its state variables, U has a certain value:

$$U = U(T, V, \ldots) .$$

All heat engines, such as steam engines, combustion motors, or heat pumps work periodically and perform *cyclic processes*. For a cyclic process, as for every state variable or function, the changes of U during one cycle must cancel:

$$\oint dU = 0 .$$

For a general process, the amount of internal energy may change by transferring heat ΔQ or mechanical work ΔW to the system (Figure 1.2). In this way, heat or mechanical work are stored inside the system. The "storage" is then in the form of internal energy ΔU. For a cyclic process, the "storage" at the end of one cycle is in the same state as it was in the beginning.

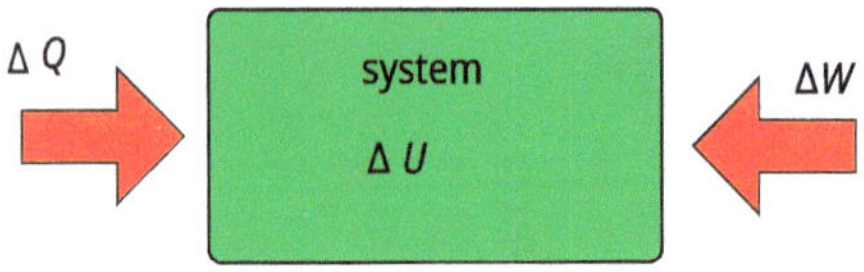

Figure 1.2: An open system may receive energy in the form of heat and/or mechanical work from its environment; both are stored in U.

1.2.2 Heat

Heat is a measure for the transfer of energy through heating, by bringing the system in thermal contact with a heat bath. The unit of heat is that of energy, Joule [J], Watt second [Ws], or calory [cal] (outdated).

- Symbol: Q;
- Transfer: ΔQ or δQ (infinitesimal);
- Def.: $\delta Q > 0$ means that heat is transferred to the system;
- Q is *not* a state function. It makes no sense to say "this state has a heat of 5 J." For a cyclic process, this implies

$$\oint \delta Q \neq 0 .$$

From our experience, we know that transferring heat to a body normally increases its temperature by $\Delta T \sim \Delta Q$, or

$$\delta Q = m \cdot c \cdot dT$$

with the specific heat c in J/(kg K). In general, c depends on temperature, pressure, etc., but also on how the process is performed:

$$\Delta Q_{12} = \int_{T_1}^{T_2} m \cdot c(T)\, dT$$

is the amount of heat needed to heat up a body with mass m from T_1 to T_2. From the above, it is clear that ΔQ_{12} will depend not only on T_1 and T_2 but also on the process itself.

1.2.3 Work

Work is a measure of the transfer of energy through changing the volume of a system by an external force F. To move the piston in Figure 1.3, the work

$$\Delta W = F \cdot \Delta x$$

is needed.

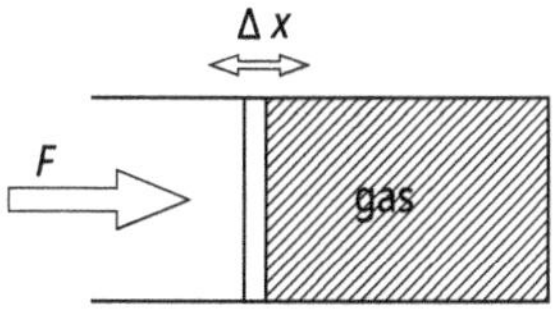

Figure 1.3: A gas in a piston receives the energy $\Delta W = F \cdot \Delta x$ if the piston moves by Δx. The pressure acting on the piston with area A is p.

The unit of work is Joule [J], Newton meter [Nm], or Watt second [Ws].
- Symbol: W;
- Transfer: ΔW or δW (infinitesimal);
- Def.: $\delta W > 0$ means that work is transferred to the system;
- W is *not* a state function. This implies

$$\oint \delta W \neq 0 .$$

1.2.3.1 Pressure

The force on the surface A is related to the pressure by

$$\vec{F} = -\int_A p\, d^2\vec{f}$$

or, infinitesimally

$$d\vec{F} = -p\hat{f}\, dA$$

with $\hat{f}$ normal to dA. Thus,

$$\delta W = d\vec{F}\, d\vec{x} = -p\, dA\, dx = -p\, dV$$

is the *work differential.* It measures the work needed to change the volume of the system by dV if it is under pressure p.

1.2.3.2 Cyclic processes

The work needed to bring a system from state 1 to state 2 in the pV-diagram, shown in Figure 1.4, depends on the path:

$$\Delta W_{12}(C_1) = -\int_{V_1, C_1}^{V_2} p\, dV = (-1) \times \text{(area under curve } C_1)$$

and

$$\Delta W_{12}(C_2) = -\int_{V_1, C_2}^{V_2} p\, dV = (-1) \times \text{(area under curve } C_2)\, .$$

Evaluating ΔW for a closed clockwise cycle yields

$$\Delta W = \Delta W_{12}(C_1) - \Delta W_{12}(C_2) = -A_C,$$

where A_C is the area inside the cycle.

1.2.4 The first law of thermodynamics

Conservation of energy leads to the global balance (Figure 1.2)

$$\Delta U = \Delta Q + \Delta W$$

or, infinitesimally

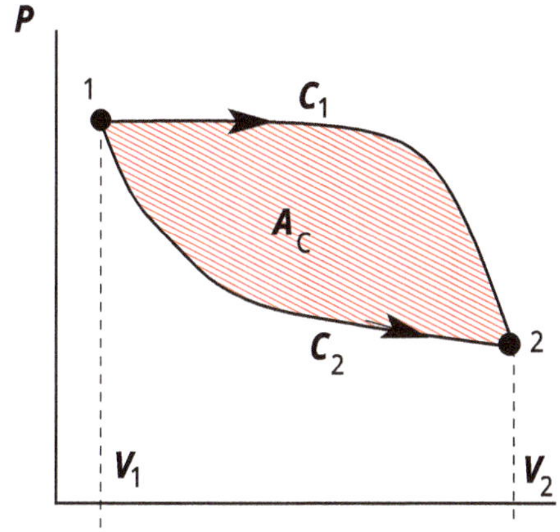

Figure 1.4: For processes connecting 1 and 2, the work transferred to or from the system depends on the path and is equal to the area under each path. If the process is cyclic and closed clockwise, the system at each cycle transfers work $\Delta W_e = A_C$ to the environment.

$$\boxed{dU = \delta Q + \delta W} \tag{1.2}$$

This is the first law of thermodynamics.

Note the different notations for infinitesimals δ and d. Since U is a state function, dU is a complete differential, but δQ, δW denote the amount transferred in/from the system and *not* the change of Q or W, because Q and W do not exist as state functions. In general for a cyclic process, we have

$$\oint dU = \underbrace{\oint \delta Q}_{\Delta Q \neq 0} + \underbrace{\oint \delta W}_{\Delta W \neq 0} = 0 \,.$$

Although both ΔQ and ΔW depend on the paths in phase space, their sum ΔU does not. Moreover, for a closed system, the first law says

$$dU = 0,$$

and thus $U = \text{const.}$ for every possible process.

The form (1.2) is also called *the equivalence of heat and work*. It shows that both are forms of energy, and that they can be transformed into each other or stored into U. However, we shall see in the following section that this kind of transfer is only possible within certain restrictions.

1.3 Entropy and the second law

1.3.1 Irreversibility and the second law with words

An everyday observation was turned into a "law" first by Clausius[1] in 1850:

1 Rudolf Clausius, German physicist, 1822–1888.

A cooler object can never heat a hotter object.

In other words, heat is always transferred from the hotter body to the colder one; see Figure 1.5.

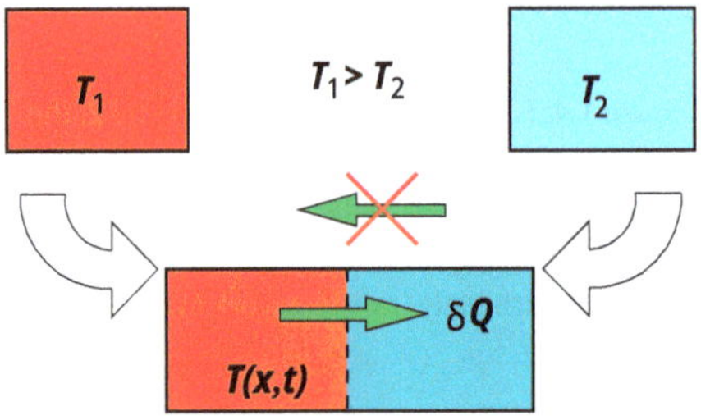

Figure 1.5: An irreversible process. The heat flows always from the warmer body to the colder one; the other direction violates the second law. At $t \to \infty$, $\delta Q = 0$, and $T(x,t) = T_0$ everywhere with $T_2 < T_0 < T_1$.

Another formulation involves heat reservoirs: *A process whose only result is to cause one heat reservoir with temperature T_2 to lose energy ΔQ through cooling and a hotter heat reservoir with $T_1 > T_2$ to gain the same amount of energy ΔQ through heating can never occur.*

Lord Kelvin's[2] formulation (1851): *A heat engine whose only result is to cause a single heat reservoir to lose energy ΔQ through cooling and perform an equal amount of work $\Delta W = \Delta Q$ is impossible.*

The second law fixes the time direction (see Figure 1.1). A backwards proceeding film showing an irreversible process would make no sense. One can also say reversing (i. e., running backwards) an irreversible process is excluded by the second law.

1.3.2 Entropy

To find a measure for irreversibility, a new state function named *entropy* is introduced. The entropy takes a unique value for a given state, independent from its history. By definition, entropy of a closed system increases if an irreversible process takes place; otherwise it is constant. Loosely speaking, the more irreversible a process is, the higher is the increment per time of the entropy. Then the second law says that in a closed system the total entropy can only increase or remain constant, no matter what kind of processes are going on. If the entropy remains constant, the system is either in equilibrium (no processes at all) or "running" processes are reversible (this is a kind of oxymoron, because reversible processes are "running" infinitely slowly and do not occur in nature). It then follows that the entropy takes a maximum if thermal equilibrium is reached. Note that this is the first place where time plays a quantitative role.

2 William Thomson, 1st Baron Kelvin, British physicist, 1824–1907.

Normally entropy is denoted by the letter S. To derive a formula for the entropy increment during a process, one needs only a few assumptions:

- S is a state function.
- S is extensive; S of a composite system is the sum of its subsystems (if they do not interact).
- S is expressed by heat and temperature. If a heat reservoir is heated, its entropy increases and vice versa.

Consider a process where heat ΔQ is exchanged at the constant temperature T. Thus, one may write

$$\Delta S = f(T, \Delta Q) \,. \tag{1.3}$$

If T is changing during the process, (1.3) is still valid for infinitesimal changes dS, δQ. Now assume that a body receiving heat ΔQ is divided into two halves, where each half receives $\Delta Q/2$. The entropy gain must still be the same and

$$\Delta S = f(T, \Delta Q) = \Delta S_1 + \Delta S_2 = f(T, \Delta Q/2) + f(T, \Delta Q/2) = 2f(T, \Delta Q/2)$$

or, more general, dividing into n pieces:

$$f(T, \Delta Q) = n\, f(T, \Delta Q/n) \,.$$

This can be only fulfilled by a linear dependence $f = g(T)\Delta Q$, and thus

$$\Delta S = g(T)\Delta Q$$

or, infinitesimally

$$dS = g(T)\delta Q \,. \tag{1.4}$$

For the time being, $g(T)$ is an arbitrary non-negative function. However, if we compute the entropy change of a hotter reservoir (T_1) giving heat $-\Delta Q$ to a colder one (T_2), it reads

$$\Delta S = [g(T_2) - g(T_1)]\Delta Q, \quad \Delta Q > 0 \,.$$

Clearly this process is irreversible, and $\Delta S > 0$. Since $g(T_2) > g(T_1)$ but $T_1 > T_2$, $g(T)$ must have a negative slope for all T.

To finally determine $g(T)$, it is sufficient to require S as a state function. Since (1.3) should not depend on the material, we may consider a perfect gas. The first law yields (1 mole):

$$\delta Q = c_V dT + \frac{RT}{V}\, dV,$$

and with (1.4)

$$dS = c_V g(T)dT + g(T)\frac{RT}{V}\,dV\,. \tag{1.5}$$

If S is a state function, (1.5) must be a complete differential, and

$$\left(\frac{\partial S}{\partial T}\right)_V = c_V g(T), \quad \left(\frac{\partial S}{\partial V}\right)_T = g(T)\frac{RT}{V}\,.$$

Therefore (integrability condition),

$$\frac{\partial}{\partial V}(c_V g(T)) = \frac{\partial}{\partial T}\left(g(T)\frac{RT}{V}\right).$$

The left hand side is zero, thus we have

$$\frac{\partial}{\partial T}(Tg(T)) = 0$$

and

$$g(T) = \frac{k}{T}$$

with an arbitrary positive constant k that can be put to one by scaling of T. Finally, we have

$$dS = \frac{\delta Q}{T}\,. \tag{1.6}$$

Note that the factor $1/T$ renders the incomplete differential δQ into a complete one and defines a new state function $S(T, V)$.

1.3.3 Remark: incomplete differential and integrating factor

A differential given by two independent variables

$$\delta f = a(x,y)dx + b(x,y)dy \tag{1.7}$$

is incomplete if

$$\frac{\partial a}{\partial y} \neq \frac{\partial b}{\partial x}\,. \tag{1.8}$$

It can always be transformed to a complete (or total) differential by multiplying (1.7) with a function $g(x,y)$, called the integrating factor:

$$g\delta f = gadx + gbdy, \tag{1.9}$$

where $d\tilde{f} = g\delta f$ is the complete differential of the existing state function $\tilde{f}(x,y)$ if

$$\frac{\partial(ga)}{\partial y} = \frac{\partial(gb)}{\partial x} \tag{1.10}$$

holds. From (1.10), $g(x,y)$ is determined. The new state function $\tilde{f}$ is then found by integrating

$$\frac{\partial\tilde{f}}{\partial x} = ga, \quad \frac{\partial\tilde{f}}{\partial y} = gb\,. \tag{1.11}$$

Example:

$$\delta f = y^2 dx + dy$$

is incomplete. Equation (1.10) turns into

$$\frac{\partial(gy^2)}{\partial y} = \frac{\partial g}{\partial x}, \tag{1.12}$$

which can be solved by $g = g(y)$. Then,

$$\frac{d(gy^2)}{dy} = 0$$

from where

$$g(y) = \frac{1}{y^2}\,.$$

The new function $\tilde{f}$ is found from (1.11)

$$\frac{\partial\tilde{f}}{\partial x} = 1, \quad \frac{\partial\tilde{f}}{\partial y} = \frac{1}{y^2},$$

from where

$$\tilde{f}(x,y) = x - \frac{1}{y}\,.$$

One can easily verify that the solution of (1.12) is not unique. Another one reads

$$g(x,y) = \frac{1}{y^2}\exp(-k/y + kx)$$

for arbitrary k and results, of course, in a quite different state function $\tilde{f}$.

1.3.4 Computation of the entropy by a reversible equivalent process

The problem computing ΔS by the help of (1.6) is that this relation is only valid for reversible processes in thermal equilibrium. If we consider a process as shown in Figure 1.6, equilibrium is left in the moment when the barrier is removed. Then heat begins to flow and temperature becomes a function of time and space, both characteristics of non-equilibrium states. Only at the very end, after an (infinitely) long time, equilibrium is reached again. But since entropy increment (remember S is a state function) is equal to the entropy difference of the initial and of the final state and does not depend on the process between, one may construct an equivalent process but which can be performed reversible. Here we consider the backward process, Figure 1.7. In a first step, the block with uniform temperature is thermally divided into two blocks, still with the same temperature (reversible process). Then the left block is slowly heated (reversible) to temperature T_1, increasing the entropy by

$$\Delta S_1 = \frac{mc_V}{2} \int_{T_0}^{T_1} \frac{1}{T} dT = \frac{mc_V}{2} \ln\left(\frac{T_1}{T_0}\right) > 0 ,$$

where constant specific heat c_V and constant volume is assumed. As a next step, the right block is slowly cooled to T_2, realeasing reversibly the entropy:

$$\Delta S_2 = \frac{mc_V}{2} \int_{T_0}^{T_2} \frac{1}{T} dT = \frac{mc_V}{2} \ln\left(\frac{T_2}{T_0}\right) < 0 .$$

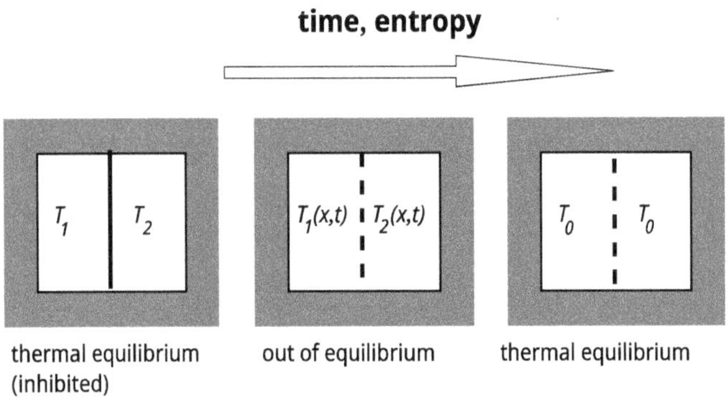

Figure 1.6: An irreversible process inside a closed system. Entropy increases from left to right as soon as the heat barrier is removed.

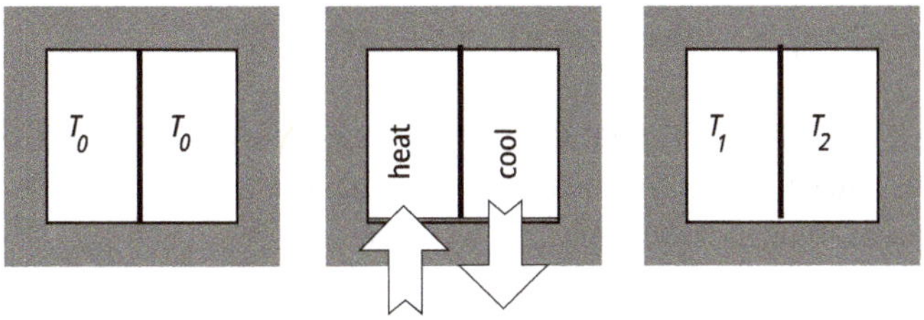

Figure 1.7: Reversible equivalent backwards process to Figure 1.6.

For the total entropy change of the reversible equivalent process, we find

$$\Delta S_e = \Delta S_1 + \Delta S_2 = \frac{mc_V}{2}\ln\left(\frac{4\,T_1T_2}{(T_1+T_2)^2}\right) = \frac{mc_V}{2}\ln\left(\frac{4\,X}{(1+X)^2}\right),$$

where we used $T_0 = (T_1 + T_2)/2$ (first law) and the abbreviation $X = T_1/T_2$. The original process goes in the other direction, thus

$$\Delta S = -\Delta S_e = \frac{mc_V}{2}\ln\left(\frac{(1+X)^2}{4\,X}\right).$$

The minimum of ΔS is at $X = 1$, corresponding to $T_1 = T_2$:

$$\Delta S_{\text{min}} = 0\,.$$

Therefore, $\Delta S > 0$ if $T_1 \neq T_2$, Figure 1.8. If $T_1 = T_2$, the process in Figure 1.6 is trivially reversible and coincides with its equivalent process (in fact, there is no process at all).

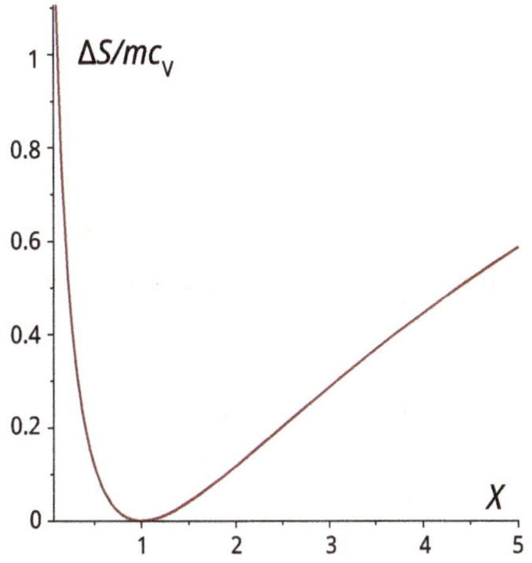

Figure 1.8: Entropy increment of the irreversible process in Figure 1.6 over $X = T_1/T_2$.

1.3.5 Entropy increase of a body in sudden contact with a heat bath

It is instructive to derive a formula for the total entropy increase if a body with mass m and temperature T comes into contact with a heat bath with mass $M \gg m$ and temperature T_b. Using the same reasoning as in Section 1.3.4, we have

$$\Delta S = c_b\, M \int_{T_b(1-\delta)}^{T_b} \frac{dT}{T} + c\, m \int_{T}^{T_b} \frac{dT}{T} = -c_b\, M \ln(1-\delta) + c\, m \ln \frac{T_b}{T} \tag{1.13}$$

with the specific heats c and c_b. The small number δ accounts for an infinitesimal change of the temperature of the heat bath and can be computed from the first law as

$$\delta = \frac{c\, m}{c_b\, M}\left(\frac{T}{T_b} - 1\right). \tag{1.14}$$

Clearly, $\delta \to 0$ for $M/m \to \infty$. Expanding the logarithm in (1.13) to the first order of δ and using (1.14), we find

$$\Delta S = c\, m\left(\frac{T}{T_b} - 1 - \ln \frac{T}{T_b}\right) \tag{1.15}$$

for the change of S of the body and the heat bath. From $\Delta S = \Delta S_{\text{min}} = 0$ for $T = T_b$ one sees that S always increases if the body had a temperature different from that of the heat bath.

1.3.6 Carnot process, entropy, and efficiency

We begin with Carnot's[3] theorem:

> *the most effective heat engine is one that operates reversibly.*

What is a heat engine, and what does "most effective" mean? A heat engine is a periodically working machine that transforms heat into mechanical work. Starting with the—now a bit out of fashion—steam engines, combustion engines, carbon, gas, oil, or nuclear feated power plants, are all heat engines. Heat pumps are invers working heat engines. A simple sketch is given in Figure 1.9. During one cycle, a heat engine extracts the energy ΔQ_1 by cooling a hotter reservoir, transforms part of this energy into work ΔW_e and rejects the rest as waste energy $-\Delta Q_2$ by heating the colder reservoir. The first law relates the energies:

3 Nicolas Léonard Sadi Carnot, French physicist, 1796–1832.

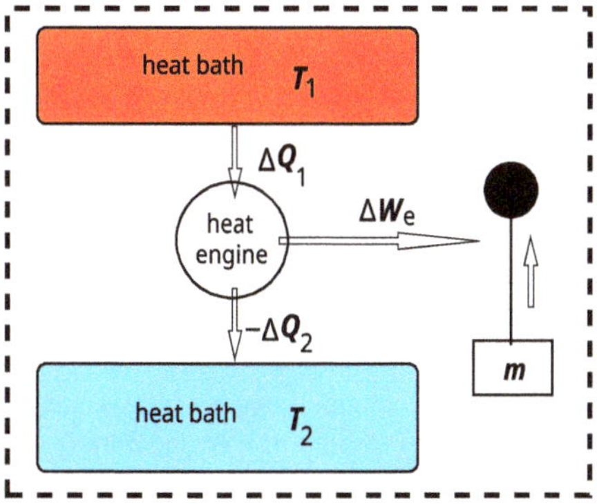

Figure 1.9: A heat engine working between two heat baths with $T_1 > T_2$ produces work that is used to lift a mass m inside the closed system.

$$\Delta U = 0 = \Delta Q_1 + \Delta Q_2 + \Delta W, \tag{1.16}$$

where $\Delta Q_1 > 0, \Delta Q_2 < 0, \Delta W = -\Delta W_e < 0$. The system is closed if the work is used inside for an idealized reversible process, i. e., lifting a weight without friction by $\Delta h = \Delta W_e/mg$ or compressing an ideal spring. For a periodically working machine, the entropy change over one cycle can take place only in the heat reservoirs:

$$\Delta S = \Delta S_1 + \Delta S_2 = -\frac{\Delta Q_1}{T_1} - \frac{\Delta Q_2}{T_2}. \tag{1.17}$$

We wish to relate entropy change to efficiency. For a heat engine, efficiency is defined as ratio of produced work ΔW_e and consumed heat:

$$\eta \equiv \frac{\Delta W_e}{\Delta Q_1} \tag{1.18}$$

with $0 \leq \eta \leq 1$ due to (1.16). We use (1.16) and (1.18) to eliminate ΔQ_1, ΔQ_2 in (1.17) and find

$$\Delta S = \frac{\Delta W_e}{T_2 \eta}\left[1 - \eta - \frac{T_2}{T_1}\right]. \tag{1.19}$$

The smaller the efficiency, the larger the entropy production per cycle and vice versa, Figure 1.10. Moreover ΔS cannot be negative in a closed system, and $\Delta S = 0$ belongs to the reversibly working engine, leading to

$$\eta \leq \eta_C = 1 - \frac{T_2}{T_1}, \tag{1.20}$$

where η_C denotes the maximal (Carnot) efficiency. The Carnot efficiency is only determined by the ratio of the two temperatures of the reservoirs.

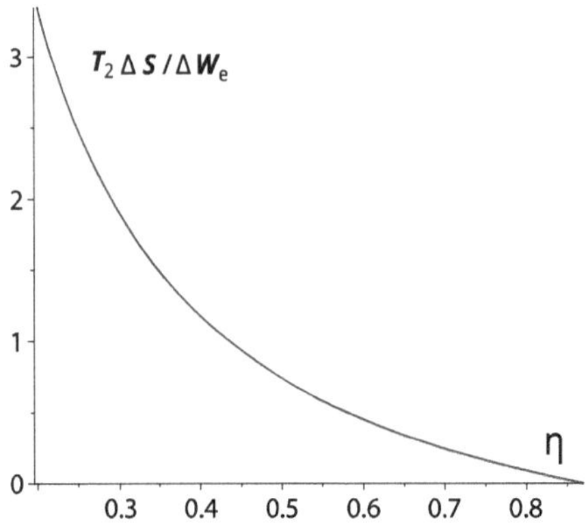

Figure 1.10: Entropy production per cycle versus efficiency of an irreversible heat engine working between $T_1 = 2300$ K and $T_2 = 300$ K. Here, $\eta_C = 0.87$.

The most effective engine is the reversible one. All reversibly working engines have the same efficiency η_C from (1.20). However, every realistic heat engine works irreversibly, and normally η is much less than η_C. Note that the important result (1.20) does not depend on the working substance of the heat engine.

1.3.7 The second law of thermodynamics

Finally, we ask again: "What is entropy?" We can now answer with the words of A. Sommerfeld:[4]

Each thermodynamic system possesses an extensive state function S called entropy. Its increase (decrease) during reversible processes is given by dividing the added (extracted) amount of heat by the absolute temperature. All irreversible processes produce entropy leading to an extra increase of S.

It can be formulated as

$$dS = \frac{\delta Q}{T} + dS_{\text{irr}} \tag{1.21}$$

with the extra irreversible amount $dS_{\text{irr}} \geq 0$, or more concisely,

$$\boxed{dS \geq \frac{\delta Q}{T}}\,. \tag{1.22}$$

This is the second law of thermodynamics.

4 Arnold Sommerfeld, German physicist, 1868–1951.

1.3.8 Heat theorem of Nernst and the third law

The heat theorem goes back to W. Nernst[5] in 1926 and claims that *the entropy of any system approaches a definite and finite value independent from its other thermodynamic variables if its temperature approaches zero*:

$$\lim_{T\to 0} S = S_0 \,.$$

It is based on empirical grounds and found its theoretical justification later in statistical thermodynamics and quantum mechanics. Planck[6] showed that $S_0 = 0$. The condition

$$\lim_{T\to 0} S = 0$$

is nowadays called the *third law of thermodynamics*. Note that the entropy for a perfect gas computed later in (1.32) will not fulfill the third law. The reason is that the EOS used to derive (1.32) are not valid for the limit $T \to 0$ and can therefore only considered as a (good) approximation for larger T.

1.4 Gibbs fundamental equation

The Gibbs fundamental equation results from a combination of the first and the second law. For reversible processes, one eliminates δQ between (1.22) and (1.2) and obtains

$$dS = \frac{1}{T}dU + \frac{p}{T}dV \,. \tag{1.23}$$

A more general form is obtained by introducing into the first law the work differentials

$$-pdV, \quad \vec{H}d\vec{M}, \quad \vec{E}d\vec{P}, \quad \text{etc.}$$

for magnetic or electric materials. They all have the form

$$a_i dA_i,$$

where a_i are intensive and A_i extensive variables. Then,

$$\boxed{dS = \frac{1}{T}dU - \frac{1}{T}\sum_i^{\ell} a_i dA_i} \tag{1.24}$$

is *Gibbs fundamental equation*, written down the first time by J. W. Gibbs[7] in 1873.

5 Walther Nernst, German physicist, 1864–1941.

6 Max Planck, German physicist, 1858–1947.

7 Josiah Willard Gibbs, American physicist, 1839–1903.

1.4.1 Thermodynamic potentials

Interpreting (1.24) as the complete differential of the state function S, it follows that

$$S = S(U, A_1 \dots A_\ell)\,.$$

The variables U, A_i are called *natural variables* of S. We can also solve (1.24) for dU,

$$dU = TdS + \sum_i^\ell a_i dA_i \tag{1.25}$$

and may consider

$$U = U(S, A_1 \dots A_\ell)$$

with the natural variables of U being S, A_i. On the other hand, comparing (1.24) with the complete differential of dS, one finds the relations

$$\frac{1}{T} = \frac{\partial S}{\partial U} \quad \frac{a_i}{T} = -\frac{\partial S}{\partial A_i}, \qquad i = 1 \dots \ell\,. \tag{1.26}$$

Solving the first one for U yields

$$U = U(T, A_i)\,, \tag{1.27}$$

the caloric state equation. The latter are solved for a_i, and with (1.27)

$$a_i = a_i(U, A_i) = a_i\big(U(T, A_i), A_i\big) = \tilde{a}_i(T, A_i) \tag{1.28}$$

one arrives at the thermal state equations. As a quintessence, we see that if S is known in its natural variables, all equations of state (EOS) can be derived. This is why $S(U, A_i)$ is called a *thermodynamic potential.* It is important to note that other representations, e. g., $S(T, A_i)$ or $S(p, A_i)$ do not include the full information, and the EOS cannot be derived completely.

The same can be performed with (1.25), yielding

$$T = \frac{\partial U}{\partial S} \quad a_i = \frac{\partial U}{\partial A_i}, \qquad i = 1 \dots \ell\,, \tag{1.29}$$

and from there

$$T = T(S, A_i) = T\big(S(U, A_i), A_i\big) = \tilde{T}(U, A_i) \tag{1.30}$$

and

$$a_i = a_i(S, A_i) = a_i\big(S(U, A_i), A_i\big) = \tilde{a}_i(U, A_i) = \tilde{a}_i\big(U(T, A_i), A_i\big) = \bar{a}_i(T, A_i)\,. \tag{1.31}$$

The first one is the caloric EOS (after solving for U), the second ones are the thermal EOS. Like $S(U, A_i)$, the state function $U(S, A_i)$ contains as well the full information of the system, namely the EOS, and is therefore also a thermodynamic potential.

1.4.2 Equations of state for a perfect gas

Let us go back to a simple system described by only two state variables and the form (1.23). As an example, we take a perfect gas.

For one mole of a perfect gas, the entropy reads

$$S(U, V) = S_0 + c_V \ln(U/U_0) + R \ln(V/V_0) . \tag{1.32}$$

Then

$$\frac{1}{T} = \left(\frac{\partial S}{\partial U}\right)_V = \frac{c_V}{U}$$

and

$$\frac{p}{T} = \left(\frac{\partial S}{\partial V}\right)_U = \frac{R}{V} .$$

On the other hand, U in its natural variables is given as

$$U(S, V) = U_0 \left(\frac{V}{V_0}\right)^{R/c_v} e^{\frac{S-S_0}{c_V}} . \tag{1.33}$$

Inserting this into (1.30) yields

$$T = \frac{U}{c_V} , \tag{1.34}$$

the caloric EOS; inserting into (1.31) gives

$$p = \frac{RU}{c_V V} \tag{1.35}$$

and, with (1.34) the thermal EOS.

1.4.3 Integrability

The EOS are related to each other. Taking $U = U(T, V)$, (1.23) reads

$$dS = \frac{1}{T}\left(\frac{\partial U}{\partial T}\right)_V dT + \frac{1}{T}\left[\left(\frac{\partial U}{\partial V}\right)_T + p\right] dV .$$

From $S(T,V)$, it follows

$$\left(\frac{\partial S}{\partial T}\right)_V = \frac{1}{T}\left(\frac{\partial U}{\partial T}\right)_V, \quad \left(\frac{\partial S}{\partial V}\right)_T = \frac{1}{T}\left[\left(\frac{\partial U}{\partial V}\right)_T + p\right].$$

Differentiating the left equation by V, the right one by T, and subtracting yields the integrability condition

$$\left(\frac{\partial U}{\partial V}\right)_T = T\left(\frac{\partial p}{\partial T}\right)_V - p \tag{1.36}$$

for the perfect gas.

Example 1: mechanical spring. A spring with spring constant D and equilibrium length h_0 is fixed inside an evacuated piston with cross section area A and a vertically movable lid (Figure 1.11). The system is in mechanical equilibrium if the force

$$F = D(h_0 - h)$$

is applied on the top lid. With $V = Ah$ and $p = F/A$, this is the thermal EOS

$$p = \tilde{D}(V_0 - V), \quad \tilde{D} = \frac{D}{A^2}.$$

From (1.36), we immediately find

$$U(T,V) = -\int p\, dV = \frac{\tilde{D}}{2}(V_0 - V)^2 + f(T)$$

with an undetermined function $f(T)$. The first part is the well-known mechanical potential energy of a spring; the latter reflects thermal properties, for which more informa-

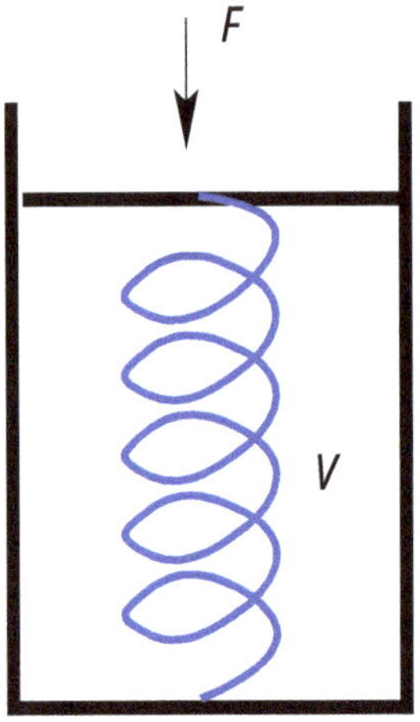

Figure 1.11: A spring inside an evacuated piston.

tion is needed (heat capacity of the spring). If, however, the spring becomes softer when temperature increases, i. e.,

$$\tilde{D} = D_0 - D_T(T - T_0), \quad D_T > 0\,,$$

the integration of (1.36) yields

$$U(T,V) = \frac{D_0 - D_1 T_0}{2}\,(V_0 - V)^2 + f(T).$$

Example 2: real gas. The thermal EOS for a real gas can be written as virial expansion:

$$p(V,T) = \frac{RT}{V}\left(1 + \frac{a_1(T)}{V} + \frac{a_2(T)}{V^2} + \cdots\right)$$

with the virial coefficients $a_i(T)$. Inserting in (1.36) gives

$$\left(\frac{\partial U}{\partial V}\right)_T = \frac{RT^2}{V}\left(\frac{a_1'(T)}{V} + \frac{a_2'(T)}{V^2} + \cdots\right),$$

where primes denote derivatives with respect to T. Integrating over V yields

$$U(V,T) = c_V T - RT^2\left(\frac{a_1'(T)}{V} + \frac{1}{2}\,\frac{a_2'(T)}{V^2} + \cdots\right),$$

where the first term on the r. h. s. is the integration constant. Contrary to the perfect gas, for a real gas the inner energy depends on the volume due to internal forces (attractive or repelling) between the molecules, reflected by a_i. Since $1/V \sim \rho$ (density), one may write U as an expansion with respect to density

$$U(\rho,T) = c_V T - RT^2 \sum_n b_n(T)\rho^n$$

with other coefficients b_i that follow from a_i.

1.5 Multi-component systems

We extend the treatment to systems with more than one component. The components can mix in arbitrary ratios; their mole numbers occur as new state variables

$$n_i, \quad n = 1\ldots k,$$

k being the number of components. As a consequence of mass conservation, the n_i may change for two reasons only: (i) in an open system by a mass flux through the borders, (ii) due to chemical reactions.

1.5.1 Open system with respect to mole number, one component

The mole number n is a new independent variable. The extensive state function U is additive:

$$U = n \cdot u,$$

where u is the molar internal energy. If n can change, the first law must be extended. For one mole, we still have

$$du = \delta q + \delta w = Tds - pdv\,, \tag{1.37}$$

and therefore

$$ndu = nTds - npdv\,.$$

From there,

$$d(nu) - udn = Td(ns) - Tsdn - pd(nv) + pvdn\,.$$

For the extensive variables, we use $S = ns$, $V = nv$ and $U = nu$, and find

$$dU = TdS - pdV + (u - Ts + pv)dn = TdS - pdV + gdn\,,$$

where

$$g = u - Ts + pv \tag{1.38}$$

is the free molar enthalpy (molar Gibbs free energy). The internal energy now becomes a function of V, S, n, and from the complete differential one concludes

$$g = \left(\frac{\partial U}{\partial n}\right)_{S,V}\,. \tag{1.39}$$

This gives a natural explanation for the molar Gibbs free energy. It is just the amount in which U in its natural variables (S, V, n) increases, if one mole is added to the system, but S and V are kept constant.

1.5.2 First law for a mixture of *k* components

For k components, one finds $U(S, V, n_1 \ldots n_k)$ and the first law takes the form

$$dU = TdS - pdV + \sum_i^k \mu_i dn_i, \tag{1.40}$$

where

$$\mu_i = \left(\frac{\partial U}{\partial n_i}\right)_{S,V}$$

is called *chemical potential* of component i. It measures the amount of energy gained (or needed) if one mole of component i is added to the system. IF U is given in its natural variables, we have

$$\mu_i = \mu_i(S, V, n_1 \dots n_k)\,.$$

Example: chemical potential of a perfect gas. If $k = 1$ (one component only), we see from (1.39) that $\mu = g$. From (1.38) and the 1st law (1.37),

$$\begin{aligned} dg &= du - Tds - sdT + pdv + vdp = Tds - pdv - Tds - sdT + pdv + vdp \\ &= -sdT + vdp, \end{aligned}$$

one finds

$$\left(\frac{\partial g}{\partial T}\right)_p = -s, \quad \left(\frac{\partial g}{\partial p}\right)_T = v\,. \tag{1.41}$$

For a perfect gas one has

$$s(p, T) = c_p \ln\left(\frac{T}{T_0}\right) - R\ln\left(\frac{p}{p_0}\right) + s_0$$

and

$$v(p, T) = \frac{RT}{p}\,.$$

Integrating (1.41) yields

$$\mu(p, T) = \mu_0 + c_p(T - T_0) - c_pT\ln\left(\frac{T}{T_0}\right) + RT\ln\left(\frac{p}{p_0}\right), \tag{1.42}$$

the chemical potential of a perfect gas.

1.5.3 Other thermodynamic potentials

Depending on the variables appropriate for a certain problem, it can be of advantage to use other potentials than U or S. A mathematical method, the *Legendre transformation* [1], allows for introducing other independent variables and to derive new potentials in their natural variables. We mention the free energy

$$F = U - TS$$

and

$$dF = -SdT - pdV + \sum_i^k \mu_i dn_i \,, \tag{1.43}$$

where the natural variables are now V, T, n_i,

$$F = F(V, T, n_i)$$

and

$$\mu_i = \left(\frac{\partial F}{\partial n_i}\right)_{T,V} . \tag{1.44}$$

The free energy F is the Legendre transform of U, and

$$\left(\frac{\partial F}{\partial T}\right)_{n_i,V} = -S$$

holds. Another important potential is the *Gibbs free energy*

$$G = F + pV = U - TS + pV$$

and

$$dG = -SdT + Vdp + \sum_i^k \mu_i dn_i \tag{1.45}$$

in its natural variables

$$G = G(p, T, n_i)$$

and

$$\mu_i = \left(\frac{\partial G}{\partial n_i}\right)_{T,p} . \tag{1.46}$$

Since G is extensive, it must increase linearly with n_i (all other variables of G are intensive):

$$G(p, T, \alpha n_i) = \alpha G(p, T, n_i) \,.$$

This must hold for all α, and we may differentiate the equation with respect to α:

$$\sum_j^k \frac{\partial G}{\partial(\alpha n_j)} \frac{\partial(\alpha n_j)}{\partial \alpha} = \sum_j^k \frac{\partial G}{\partial(\alpha n_j)} n_j = G(p, T, n_i) \,.$$

For $a = 1$ and with (1.46),

$$G(p, T, n_i) = \sum_j^k n_j \, \mu_j(p, T, n_i) \,. \tag{1.47}$$

This is the important *Gibbs–Duhem relation.*

On the other hand, any intensive quantity, such as μ_i, must not depend on the system size,

$$\mu_j(p, T, an_i) = \mu_j(p, T, n_i) \,.$$

Derivation with respect to a on both sides leads to

$$\sum_i^k n_i \frac{\partial \mu_j}{\partial n_i} = 0 \,. \tag{1.48}$$

Inserting (1.46) yields

$$\sum_i^k n_i \frac{\partial^2 G}{\partial n_i \partial n_j} = \sum_i^k n_i \frac{\partial \mu_i}{\partial n_j} = 0 \,. \tag{1.49}$$

Equations (1.48) and (1.49) are named the *Duhem–Margule relations.* They can be formulated for every intensive quantity that depends only on intensive variables (such as p, T), but also on the n_i.

For more details on thermodynamic potentials and Legendre transforms see textbooks, e. g., [1].

1.5.4 Mixture of *k* perfect gases

We consider first two separated perfect gases (n_1, n_2 mole) with the volumes $V_1/V = n_1/n$, $V_2/V = n_2/n$ and $V = V_1 + V_2$, $n = n_1 + n_2$. Assume they have before mixing the same temperature T and the same pressure p, Figure 1.12. After removing the wall, the gases will mix homogeneously, where temperature and pressure remain unchanged. This is definitely an irreversible process, and we want to compute the increase of entropy.

In the beginning we have two independent systems with

$$S_b = n_1 s(T, V_1) + n_2 s(T, V_2), \tag{1.50}$$

where $s(T, V)$ denotes the molar entropy of a perfect gas:

$$s(T, V) = s_0 + c_V \ln\left(\frac{T}{T_0}\right) + R \ln\left(\frac{V}{V_0}\right) . \tag{1.51}$$

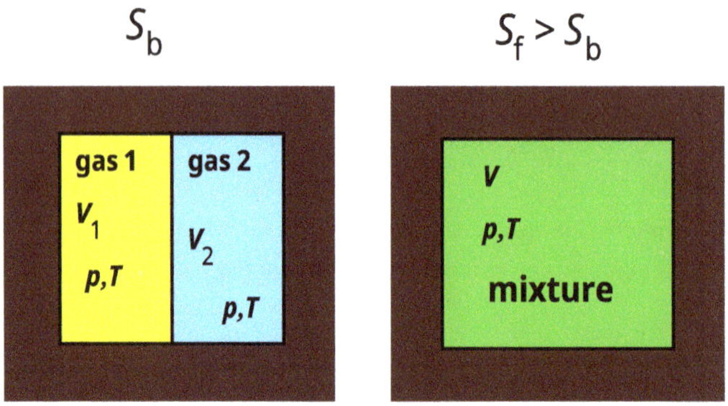

Figure 1.12: Mixing of two gases inside a closed system is an irreversible process, where the total entropy must increase.

The final state has the entropy

$$S_f = ns(T, V) .$$

From here, one computes

$$\Delta S = S_f - S_b = R\, n_1 \ln\left(\frac{n}{n_1}\right) + R\, n_2 \ln\left(\frac{n}{n_2}\right) > 0 \tag{1.52}$$

as the entropy of mixing. Finally, this can be generalized to k cells with k perfect gases, resulting in

$$\Delta S = \sum_i^k R\, n_i \ln\left(\frac{n}{n_i}\right) > 0 \tag{1.53}$$

with $n = \sum_i n_i$.

1.5.5 Entropy and Gibbs free energy for a mixture of perfect gases

From the first law, we can express ds (molar) as

$$ds = \frac{du}{T} + \frac{p}{T} dv .$$

For a perfect gas

$$du = c_V dT, \quad dv = (\partial v/\partial T)dT + (\partial v/\partial p)dp = (R/p)dT - (RT/p^2)dp$$

and

$$ds = \frac{c_p}{T} dT - \frac{R}{p} dp$$

with $c_p = R + c_v$ giving the representation $s = s(T, p)$ and

$$\left(\frac{\partial s}{\partial T}\right)_p = \frac{c_p}{T}, \quad \left(\frac{\partial s}{\partial p}\right)_T = -\frac{R}{p}.$$

Integrating the last two conditions yields

$$s(p, T) = c_p \ln\left(\frac{T}{T_0}\right) - R \ln\left(\frac{p}{p_0}\right) + s_0 \tag{1.54}$$

for the entropy of one mole of a perfect gas. For a mixture of k perfect gases with n_j mole, we receive

$$S(p, T, n_i) = \sum_j^k n_j \left(c_p^{(j)} \ln\left(\frac{T}{T_0}\right) - R \ln\left(\frac{p_j}{p_0}\right) + s_0 \right), \tag{1.55}$$

where we allow for different heat capacities $c_p^{(j)}$ and p is replaced by the partial pressure

$$p_j = \frac{n_j RT}{V} = \frac{n_j}{n} p. \tag{1.56}$$

Why? One can do the following "Gedankenexperiment": take k gases with mole number n_i, each gas in a separate volume V. Each gas is then under the partial pressure p_i. Now bring them together in one volume V. If the gases have no interaction (ideal gases) the entropy of the mixture is given by (1.55), which reads with (1.56)

$$S(p, T, n_i) = \sum_j^k n_j \left(c_p^{(j)} \ln\left(\frac{T}{T_0}\right) - R \ln\left(\frac{p}{p_0}\right) + s_0 \right) + R \sum_j^k n_j \ln\left(\frac{n}{n_j}\right). \tag{1.57}$$

In the last term, we easily identify the entropy of mixing (1.53).

Computing Gibbs free energy, we use the Legendre transform

$$G = U - TS + pV = \sum_j^k n_j u_j - TS + pV$$

with $u_j = c_V^{(j)} T$ (perfect gas). Inserting S from (1.57), G can be written as

$$G(p, T, n_i) = \sum_j^k n_j \left(g_j(p, T) - RT \ln\left(\frac{n}{n_j}\right) \right), \tag{1.58}$$

where

$$g_j(p,T) = c_p^{(j)}T - c_p^{(j)}T\ln\left(\frac{T}{T_0}\right) + RT\ln\left(\frac{p}{p_0}\right) - Ts_0$$

as the molar Gibbs free energy of component j (we used $c_p^{(j)} - c_V^{(j)} = R$). Comparing (1.58) with the Gibbs–Duhem relation (1.47), one finds for the chemical potentials of each component j

$$\mu_j(p,T,n_i) = g_j(p,T) + RT\ln\left(\frac{n_j}{n}\right). \tag{1.59}$$

For one component $k = 1$, this reduces to (1.42).

1.5.6 Gibbs fundamental equation

From (1.40), Gibbs fundamental equation for a k-component mixture follows, which reads in the more general form of (1.24)

$$\boxed{dS = \frac{1}{T}dU - \frac{1}{T}\sum_i^{\ell} a_i dA_i - \frac{1}{T}\sum_j^{k} \mu_j dn_j}\,. \tag{1.60}$$

1.5.7 Equilibrium conditions

From the 2nd law, we know that for real processes going on in a closed system ($dU = 0$, $dV = 0$, $dn_i = 0$) the entropy increases and finds its maximum in thermal equilibrium. But very often, one has to deal with open systems, where for instance pressure and temperature are controlled from outside. This may be the case for phase transitions or for chemical reactions.

It is convenient to consider the Gibbs free energy if T and p are controlled. Taking again the Legrendre transform

$$G = U - TS + pV$$

and the first law (irreversible formulation with δQ)

$$dU = \delta Q - pdV + \sum_i^k \mu_i dn_i\,,$$

we find

$$dG = \delta Q - TdS - SdT + Vdp + \sum_i^k \mu_i dn_i\,. \tag{1.61}$$

Now we express δQ by the 2nd law in the form of (1.21)

$$\delta Q = TdS - TdS_{\rm irr}$$

with $dS_{\rm irr} > 0$ if irreversible processes are taking place. Finally, we have from (1.61)

$$dG = -SdT + Vdp + \sum_i^k \mu_i dn_i - TdS_{\rm irr}, \qquad (1.62)$$

and therefore $dG \le 0$ if T, p and the mole numbers are kept constant. Thus, G decreases as long as irreversible processes are at work and reaches its minimum in equilibrium. The equilibrium condition reads

$$dG^{\rm eq} = 0, \quad G^{\rm eq} = G_{\rm min}\,. \qquad (1.63)$$

1.5.8 Chemical reactions

In a closed system with impermeable walls, the mole numbers n_i can only vary due to chemical reactions. Consider a reaction in its general form

$$\sum_{i=1}^{L}(-\nu_i)B_i \rightleftharpoons \sum_{i=L+1}^{k} \nu_i\, B_i\,, \qquad (1.64)$$

where ν_i are the stoichiometric coefficients and B_i denote the reactants. As an example, we mention the reaction

$$2\,H_2 + O_2 \rightleftharpoons 2\,H_2O,$$

which will be considered later in more detail. We may assign the mole numbers

$$H_2 \rightarrow n_1, \quad O_2 \rightarrow n_2, \quad H_2O \rightarrow n_3\,,$$

and the stoichiometric coefficients would read

$$\nu_1 = -2, \quad \nu_2 = -1, \quad \nu_3 = 2\,.$$

The mole numbers cannot vary independently but are linked through the so-called extent of reaction ξ via

$$n_i = n_i^{(0)} + \nu_i\,\xi, \qquad (1.65)$$

where $n_i^{(0)}$ denote the initial values of the mole numbers. If temperature and pressure are assumed being constant, an equilibrium condition is computed from (1.63) with (1.62) for

$$dG^{eq} = \sum_i^k \mu_i dn_i = 0\,.$$

Note that in equilibrium, $dS_{irr} = 0$. Inserting (1.65), one has $\sum_i^k \mu_i \nu_i d\xi = 0$, and finally the relation

$$\sum_i^k \mu_i \nu_i = 0 \tag{1.66}$$

for the equilibrium mole numbers.

For an ideal gas mixture, we may use the chemical potentials from (1.59), and (1.66) turns into

$$\sum_i^k \nu_i \left[\frac{g_i(p,T)}{RT} + \ln X_i\right] = 0,$$

where we have introduced the mole fractions $X_i = n_i/n$. This can be written as

$$\prod_i^Z X_i^{\nu_i} = \exp\left[-\sum_i^k \frac{\nu_i g_i(p,T)}{RT}\right] = K(p,T)\,. \tag{1.67}$$

The last relation is called *the law of mass action* and goes back to experimental work of Maximilian Guldberg and Peter Waage in 1864. In its more explicit form

$$K = \frac{X_{L+1}^{\nu_{L+1}} X_{L+2}^{\nu_{L+2}} \cdots X_k^{\nu_k}}{X_1^{|\nu_1|} X_2^{|\nu_2|} \cdots X_L^{|\nu_L|}},$$

it becomes clear that the magnitude of K determines if the concentration of the reactants on the left hand side (small K) or those of the right hand side (large K) of the reaction equation prevail in equilibrium.

The law of mass action can also be formulated with the partial pressures (1.56). Substituting $X_i = p_i/p$ in (1.67) yields

$$\prod_i^Z p_i^{\nu_i} = K_p(T), \tag{1.68}$$

where the new constant

$$K_p(T) = p^{\sum_i \nu_i} K(p,T) \tag{1.69}$$

depends only on temperature and not on pressure (see Problem 1.6.9).

1.6 Problems

1.6.1 Internal energy

Consider a room with permeable but thermally isolated walls. Assume the air inside and outside the room having equal pressure (mechanical equilibrium). Consider the air as a perfect gas.

a) Show that while heating the room the internal energy of the room remains unchanged.
b) Compute the change of the internal energy of the environment due to heating of the room from T_0 to T_1. Assume that the temperature of the environment remains constant (heat bath) at $T = T_0$.
c) If you bring a cold bottle of wine from your cellar in the warm room and let it heat up, where does the energy come from?

1.6.2 Work

a) Compute the work necessary to compress a perfect gas in a cylinder to the half of its initial volume if its temperature is kept constant (isothermal process).
b) Compute the same work for a thermally isolated cylinder (adiabatic process). How large is the final temperature?

1.6.3 Processes

A thermally isolated cylinder contains n mole of a perfect gas with temperature T_0 and pressure p_0. By heating, the gas receives the heat Q_{ext}.

a) Compute temperature and pressure T_1 and p_1 after heating.
b) Now assume that the top of the cylinder can move without friction (ideal piston). Compute temperature and volume after heating.

1.6.4 Work again

Repeat Problem 1.6.2, but for one mole of a van der Waals gas.

Hints:

1) The EOS for a van der Waals gas reads (one mole) $(p + \frac{a}{V^2})(V - b) = RT$.
2) You will need the relation $(\frac{\partial U}{\partial V})_T = T(\frac{\partial p}{\partial T})_V - p$, derived in Section 1.4.3.

1.6.5 Entropy

a) Compute the entropy $S(T, V)$ for one mole of a perfect gas using Gibbs fundamental equation and the thermal and caloric EOS.
b) Now compute the entropy $S(T, V)$ for one mole of a van der Waals gas.
Hint:
The caloric EOS of a van der Waals gas reads (one mole)

$$U(T, V) = c_V T - \frac{a}{V} \,.$$

Why?

1.6.6 Complete differentials

Which of the following are complete differentials? If possible, determine $f(x,y)$.

$$\begin{aligned} df &= e^x dx + e^y dy, \\ df &= e^y dx + e^x dy, \\ df &= e^{(x+y)}(dx + dy), \\ df &= (x-y)^2 dx + (x+y)^2 dy, \\ df &= x dx + y dy, \\ df &= y dx + x dy \,. \end{aligned}$$

1.6.7 Entropy and equations of state

For a liquid at given temperature T_0 and with volume V_0, we may approximate a linear dependence of pressure on temperature and volume for $T \approx T_0$, $V \approx V_0$ (Taylor expansion). The thermal EOS thus reads

$$p(V, T) = p_0 + c_1(T - T_0) - c_2(V - V_0) \,. \tag{1}$$

a) Express the constants c_i by the help of the thermal pressure coefficient $\beta = \frac{1}{p}(\frac{\partial p}{\partial T})_V$ and the isothermal expansion coefficient $\kappa = -\frac{1}{V}(\frac{\partial V}{\partial p})_T$.
b) The caloric EOS shall be given as

$$U(T, V) = U_0 + C_V(T - T_0) + f(V) \tag{2}$$

with $U_0 = U(T_0, V_0)$. Determine the unknown function $f(V)$ from the integrability condition by the help of the thermal EOS (1).

c) Determine the entropy $S(T, V)$ from (1) and (2).

1.6.8 Entropy and irreversibility

A weight with mass m initially located at height $h > h_0$ falls on a piston, Figure 1.13. The ideal but viscous gas (one mole) in the piston chamber is compressed from V_0 to V_1. In the long time limit (final state), $t \to \infty$, the piston is in rest again due to friction. The temperature of gas, piston, and weight shall be assumed constant during the whole process (coupling to a heat bath). Energy loss due to friction during the free fall and also due to deformation of weight and piston during the impact can be neglected.

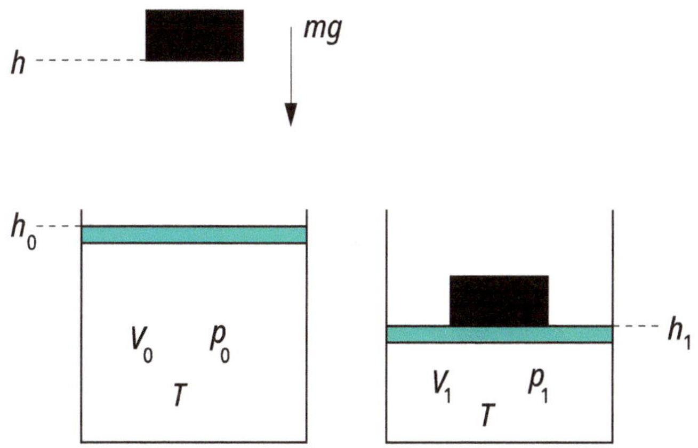

Figure 1.13: Initial state (left) and final state for $t \to \infty$ (right).

The cross-section area of the piston is given as A, so $V_0 = Ah_0$, $V_1 = Ah_1$.

a) Compute the final position h_1 of the piston as a function of m, T, and h_0.
b) Does the entropy of the system, including the heat bath, decrease, increase, or stay constant, and why?
c) Compute the entropy difference between initial and final state of the system and the heat bath as a function of m, T, h_0, and h.
d) Repeat the computation if the piston is thermally isolated. What is the final temperature of the gas?

1.6.9 Law of mass action

Show that the constant $K_p(T)$, defined in (1.69), does not depend on pressure.

Hint: compute the derivative

$$\frac{\partial K_p}{\partial p} = p^{\sum_i \nu_i - 1}\left(p\,\frac{\partial K}{\partial p} + K\sum_i \nu_i\right)$$

with K from (1.67), and then show that the bracket is zero.

2 Out of thermal equilibrium

Up to now, all state variables were homogeneous in space and constant (quasi-static) in time. But in general, quantities such as temperature, pressure, density, etc., do depend on space and time, and quite often in a very involved and irregular way, as for example in fluid dynamics of the atmosphere and the oceans. In these cases, processes are neither reversible, nor in thermal equilibrium. Fluxes such as heat and mass flows have a finite size and, as a consequence, one has to extend equilibrium thermodynamics, leading to the discipline called *Non-Equilibrium Thermodynamics* [3].

2.1 The concept of local equilibrium

2.1.1 Space and time dependent variables

As shown in Chapter 1, extensive state functions, such as energy and entropy, but also intensive state variables, for example, temperature and pressure, are well defined only in thermal equilibrium.

If the system is out of equilibrium, an anyhow generalized form of the state variables must depend on space and on time. One can still use the concept of thermodynamic state functions if one assumes local equilibrium. A system far from equilibrium can be thought divided in many small cells, the subsystems. Each subsystem, or cell, should be large enough to be treated as a thermodynamic system (many particles), but nevertheless so small that its state variables are in a good approximation homogeneously distributed over the subsystem, and that it can be considered being in equilibrium. All variables such as temperature, pressure, etc., and all state functions are defined as in equilibrium theory and are constant in each cell. But they may vary slowly from cell to cell, making them dependent on space and, due to fluxes between neighbored cells, also on time.

2.1.2 Simple networks

For illustration, let us start with the most simple case of only two cells, connected with a thermally conducting slab, Figure 2.1. Assume that both cells are in thermal equilibrium and we can define an internal energy U_i and a temperature T_i in each cell. Along the slab (not in thermal equilibrium), heat flows from the hotter (T_2) to the colder cell. We can define a heat flux density (heat per time and area)

$$j_{21}^{Q} = \frac{1}{A}\frac{\Delta Q}{\Delta t} > 0$$

with A as the cross-section area of the slab. From the first law, we find in cell 1

https://doi.org/10.1515/9783110794137-003

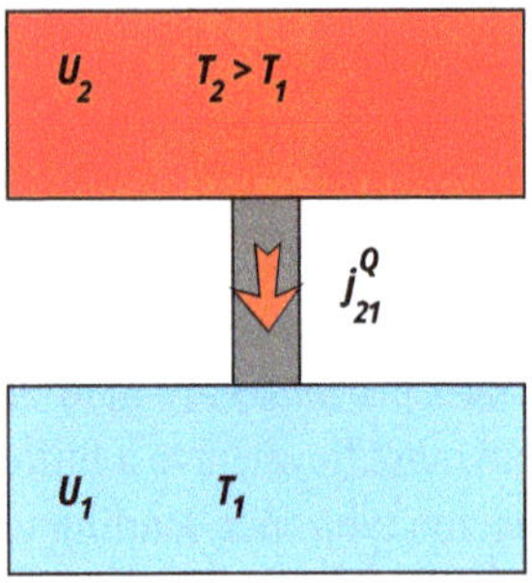

Figure 2.1: Two cells are connected by a thermally conducting slab. Both cells are in thermal equilibrium, but not the slab, where heat flows from cell 2 to cell 1.

$$\frac{\Delta U_1}{\Delta t} = \frac{\Delta Q}{\Delta t} = j_{21}^Q\, A$$

and in cell 2

$$\frac{\Delta U_2}{\Delta t} = -\frac{\Delta Q}{\Delta t} = -j_{21}^Q\, A = j_{12}^Q\, A\,.$$

For the total internal energy, we have

$$\Delta U = \Delta U_1 + \Delta U_2 = 0, \qquad U = \text{const.}\,,$$

and for the entropy from the second law

$$\Delta S = \Delta S_1 + \Delta S_2 = \Delta Q\left(\frac{1}{T_1} - \frac{1}{T_2}\right) > 0$$

or

$$\frac{\Delta S}{\Delta t} = j_{21}^Q\, A\left(\frac{1}{T_1} - \frac{1}{T_2}\right) > 0\,.$$

Entropy is produced as long as a heat flux exists, in agreement with the 2nd law, because the process is irreversible.

Now consider the slightly more complicated network in Figure 2.2. The cells are in contact with their neighbors along their edges and slabs are no longer needed. In two or three dimensions, the flux becomes a vector. The same reasoning as above applies to the energy change in cell ij,

$$\Delta V\,\frac{\Delta u_{ij}}{\Delta t} = A\,(j_{i,j}^x - j_{i+1,j}^x + j_{i,j}^y - j_{i,j+1}^y),$$

where we have introduced the energy density $u_{ij} = U_{ij}/\Delta V$ with the cell volume ΔV and A as the surface area of each cell side (we leave the superscript Q at j for a moment). Dividing by ΔV yields

$$\frac{\Delta u_{ij}}{\Delta t} = \frac{j_{i,j}^x - j_{i+1,j}^x}{d} + \frac{j_{i,j}^y - j_{i,j+1}^y}{d} \tag{2.1}$$

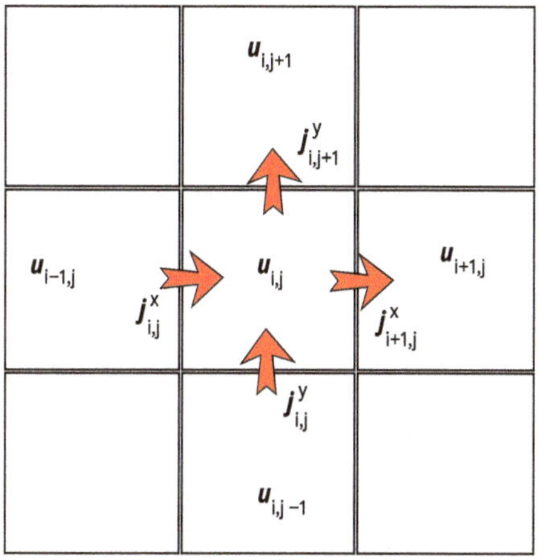

Figure 2.2: Now the cells are in thermal contact with their neighbors and heat fluxes may flow between the cells as before.

with d as the distance between two cell centers. If we identify each node ij with its location $\vec{r} = id\hat{e}_x + jd\hat{e}_y + kd\hat{e}_z$, we may write (three dimensional generalization)

$$u_{ijk} = u(\vec{r})$$

and with the limit $d \to 0$, (2.1) turns into

$$\frac{\Delta u(\vec{r})}{\Delta t} = -\nabla \cdot \vec{j}^Q(\vec{r}) \tag{2.2}$$

with the nabla operator $\nabla = (\partial_x, \partial_y, \partial_z)$. For the entropy rate, we obtain in the same way

$$\frac{\Delta s_{ij}}{\Delta t} = \frac{1}{T_{ij}} \left(\frac{j^x_{i,j} - j^x_{i+1,j}}{d} + \frac{j^y_{i,j} - j^y_{i,j+1}}{d} \right) \tag{2.3}$$

or, in the continuum limit of (2.2),

$$\frac{\Delta s(\vec{r})}{\Delta t} = -\frac{1}{T(\vec{r})} \nabla \cdot \vec{j}^Q(\vec{r}) \, . \tag{2.4}$$

Integrating (2.2) over the total volume yields for a closed system, where no heat flux exists through the surface of that volume,

$$\frac{\Delta U}{\Delta t} = 0.$$

On the other hand, integrating (2.4) gives

$$\frac{\Delta S}{\Delta t} = \int_V d^3\, \vec{r} \vec{j}^Q \cdot \nabla \frac{1}{T} \, ,$$

a quantity, that is larger zero if the heat flux points everywhere in the direction of the negative temperature gradient, which seems to be justified well by Fourier's law $\vec{j}^Q \sim -\nabla T$. As long has heat flows, the entropy increases in time, in agreement with the 2nd law.

Equations (2.2) and (2.4) have the form of balance equations, where the change of a certain density is locally connected with the sources and sinks of a flux density. In the next paragraph we shall generalize this feature.

2.1.3 Balance equations and conservation laws for densities

Consider an inhomogeneous system in local equilibrium. Let A be an extensive state function, such as energy or entropy. We assume A slowly varying (quasistatically) in time so that thermal equilibrium can still be assumed locally. Instead of A, we write now

$$A(t) = \int_V a(\vec{r}, t)\, dV,$$

where $\vec{r}$ points to the location of the subsystem, and the intensive variable $a(\vec{r}, t)$ is the density of A.

How may A change inside a given finite volume V (Figure 2.3)? We split

$$\frac{dA}{dt} = \frac{d_e A}{dt} + \frac{d_i A}{dt} = \int_V \frac{\partial a(\vec{r}, t)}{\partial t}\, d^3\vec{r} \tag{2.5}$$

where d_e denotes changes caused from outside due to currents and d_i the changes from inside due to local sources. Equation (2.5) is called a *global balance equation*. Let F be the surface of the volume V. Applying Gauss theorem, we may write for the changes from outside

$$\frac{d_e A}{dt} = -\oint_{F(V)} \vec{j}_A\, d^2\vec{f} = -\int_V \nabla \cdot \vec{j}_A\, d^3\vec{r} \tag{2.6}$$

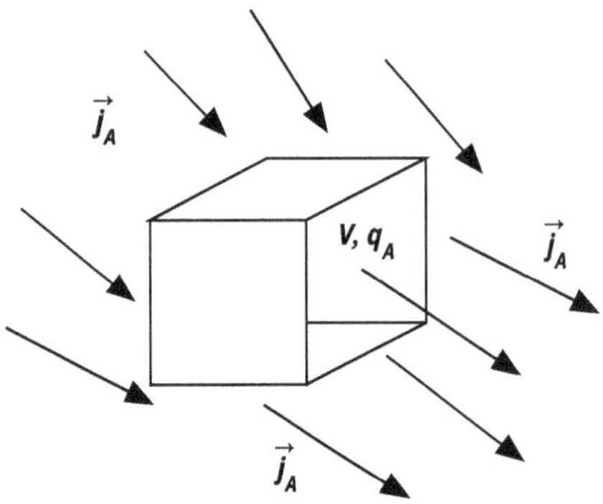

Figure 2.3: A finite volume V with sources q_A of A inside the volume and a current density $\vec{j}_A$ of A through its surface.

with the flux density $\vec{j}_A$ of A. Changes inside can be only due to sources or sinks in V

$$\frac{d_i A}{dt} = \int_V q_A \, d^3\vec{r}, \tag{2.7}$$

where q_A specifies the strength of the sources or sinks. Note that in the case of thermal equilibrium neither sources nor sinks exist and all time derivatives vanish.

Inserting (2.6), (2.7) into (2.5) and allowing V being arbitrary yields

$$\boxed{\frac{\partial a}{\partial t} + \nabla \cdot \vec{j}_A = q_A} \,. \tag{2.8}$$

This is a local balance equation. Note the similarity of (2.8) and (2.2), where we did not consider sources. Even relation (2.4) can be brought into the form of (2.8), if we write it as

$$\frac{\Delta s(\vec{r})}{\Delta t} = -\nabla \cdot \left(\frac{\vec{j}^Q}{T}\right) + \vec{j}^Q \cdot \nabla\left(\frac{1}{T}\right). \tag{2.9}$$

From here we can easily identify

$$\vec{j}^D_S = \frac{\vec{j}^Q}{T}$$

as (conductive) entropy flux (reversible part) and

$$\sigma = \vec{j}^Q \cdot \nabla\left(\frac{1}{T}\right)$$

as local entropy source (due to irreversibilty).

If the material is moving with a velocity field $\vec{v}(\vec{r}, t)$, there is an additional flux, namely the convective flux. Then the total flux of a quantity A reads

$$\vec{j}_A = \vec{v}\, a + \vec{j}^D_A, \tag{2.10}$$

where $\vec{j}^D_A$ denotes the conductive flux (or diffusive flux), in our former examples the heat or entropy flux.

2.1.3.1 Mass balance

For a, we take the mass density ρ. Mass cannot diffuse from one cell to the other; the mass flux is purely convective and reads $\vec{j}_M = \rho\,\vec{v}$. If mass is conserved (no nuclear reactions or decay, etc.), one has $q_M = 0$, and (2.8) turns into

$$\frac{\partial \rho}{\partial t} + \nabla \cdot (\rho\,\vec{v}) = 0\,. \tag{2.11}$$

This is the well-known *continuity equation*, used in fluid mechanics. In hydrodynamics, the approximation of an incompressible fluid (ρ = const.) is often good, and (2.11) reduces to

$$\nabla \cdot \vec{v} = 0 \,. \tag{2.12}$$

This is the *incompressibility condition.*

2.1.3.2 Momentum balance

The momentum density is given by $\rho\vec{v}$. Momentum changes if external forces with density $\vec{f}$ act on a subsystem, thus $\vec{q}_P = \vec{f}$. The convective transport of the i-th component reads

$$\vec{v}\rho v_i \,.$$

To complete $\vec{j}_{P_i}$ a possible conductive flux generated by friction between the subsystems has to be added:

$$\vec{j}_{P_i} = \vec{v}\rho v_i + \vec{j}^{D}_{P_i} \,.$$

Writing the three balance equations for the three components in vector notation, one finds

$$\frac{\partial(\rho\vec{v})}{\partial t} + \nabla \cdot (\rho\vec{v} \otimes \vec{v} - \underline{\tilde{\tau}}) = \vec{f}, \tag{2.13}$$

where the components of the dyadic product (tensor) read

$$(\vec{v} \otimes \vec{v})_{ij} = v_i v_j,$$

and the conductive fluxes of the three momentum components form the stress tensor

$$\tilde{\tau}_{ij} = -j^{D}_{P_i,j} \,.$$

Taking a cube oriented parallel to the coordinate axes $\hat{e}_i$ (Figure 2.4), the normal tension on each side of the cube is given by

$$\hat{e}_i \cdot \underline{\tilde{\tau}} \cdot \hat{e}_i = \tilde{\tau}_{ii} \,.$$

For a fluid in rest or for a fluid without viscosity (inviscid fluid),

$$\tilde{\tau}_{11} = \tilde{\tau}_{22} = \tilde{\tau}_{33} = -p$$

must hold, where p is the pressure inside the cube (local equilibrium). One may extract the "pressure part" from the stress tensor and write (2.13) in the form

$$\frac{\partial(\rho\vec{v})}{\partial t} + \nabla \cdot (\rho\vec{v} \otimes \vec{v} - \underline{\tau}) = \vec{f} - \nabla p, \tag{2.14}$$

where the relation

$$\tau_{ij} = \tilde{\tau}_{ij} + p\delta_{ij}$$

holds.

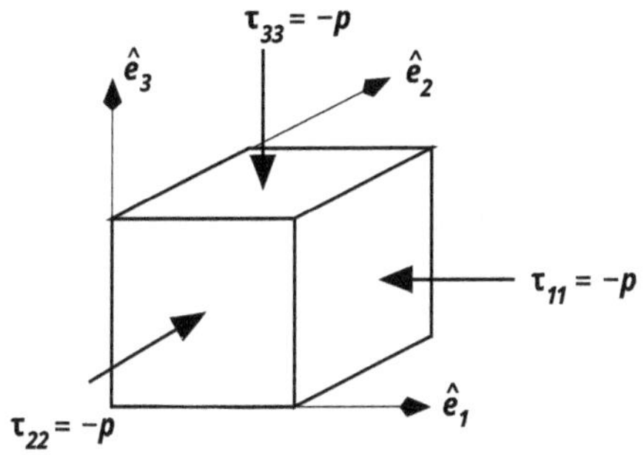

Figure 2.4: In an isotropic fluid at rest, the normal stresses on each side of an infinitesimally small cube must all be equal to the (negative) pressure inside the cube (mechanical equilibrium).

2.1.3.3 Internal energy balance

For the inner energy, we found with equation (2.2), $\vec{j}^D = \vec{j}^Q$, and then

$$\frac{\partial u}{\partial t} + \nabla \cdot (\vec{v}u + \vec{j}^Q) = q_U \tag{2.15}$$

with the local heat sources q_U. We shall see later that local heat sources may be due to viscous friction or to chemical reactions.

2.1.3.4 Entropy balance

For the entropy, we already derived (2.9), which reads with additional convective flux and $\delta t \to 0$,

$$\frac{\partial s}{\partial t} + \nabla \cdot (\vec{v}s + \vec{j}_S^D) = \sigma\,, \tag{2.16}$$

with $\vec{j}_S^D$ and σ as a generalization of the flux and source terms we found in (2.9). Their complete form will be given later.

2.1.3.5 Material derivative

If one is interested in the change of any quantity inside a volume element that moves with the material with velocity $\vec{v}(\vec{r}, t)$, one has to differentiate the density at the co-moving place $\vec{r}(t)$,

$$a(\vec{r}(t), t)$$

with respect to time. Applying the chain rule, we obtain

$$\frac{da}{dt} = \frac{\partial a}{\partial t} + \sum_i \frac{\partial a}{\partial x_i}\frac{\partial x_i}{\partial t} = \frac{\partial a}{\partial t} + \sum_i \frac{\partial a}{\partial x_i} v_i = \frac{\partial a}{\partial t} + \vec{v} \cdot \nabla a \tag{2.17}$$

with $\partial_t \vec{r}(t) = \vec{v}(\vec{r}, t)$. The expression (2.17) is called material, substantial, or Lagrangian derivative. The first term of the material derivative is the change at a fixed position (Eulerian derivative); the second one gives the changes caused by the motion of the location and is called transport term, Figure 2.5.

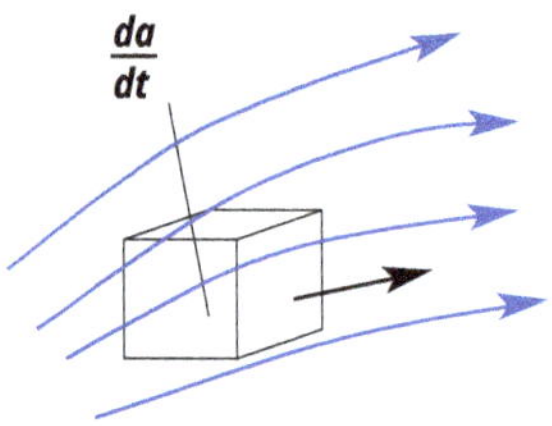

Figure 2.5: The material derivative (2.17) gives the change of a quantity at a location that moves with the material (fluid).

As an example, we take the continuity equation (2.11) and formulate it with the material derivative as

$$\frac{d\rho}{dt} = -\rho \nabla \cdot \vec{v} \,. \tag{2.18}$$

From there, it is obvious that if $\vec{v}$ has a sink (source) the density increases (decreases) in the co-moving volume element. It is also not quite correct to speak about a constant density in the case of an incompressible fluid, since with (2.12) one finds from (2.18)

$$\frac{d\rho}{dt} = 0$$

what allows for density fluctuations that move with the material.

Writing (2.15) in terms of the material derivative, we obtain

$$\frac{du}{dt} = -\nabla \cdot \vec{j}^Q + q_U - u \nabla \cdot \vec{v} \,, \tag{2.19}$$

where we recognize heat conduction and sources. The extra term $u\nabla \cdot \vec{v}$ reflects the energy change of the co-moving volume element caused by density changes. If we neglect heat conduction and heat sources for a moment, we may write

$$u = \frac{\Delta n}{\Delta V} U_0,$$

where Δn is the number of particles in ΔV, and U_0 is the energy that is contributed by one particle. Taking

$$\rho = \frac{\Delta n}{\Delta V}\, m$$

with the particle mass m, this yields

$$u = \frac{U_0}{m}\, \rho$$

and

$$\frac{du}{dt} = \frac{U_0}{m}\frac{d\rho}{dt} = \frac{u}{\rho}\frac{d\rho}{dt} = -u\nabla\cdot\vec{v}\,,$$

where we used the continuity equation (2.18). This is exactly the last term in (2.19).

2.1.4 Balance equations and conservation laws for specific quantities

Next, we resort to the specific variables, which are related to mass (or mole) instead of volume, as already defined in (1.1):

$$\hat{a}(\vec{r},t) = \frac{dA}{dM} = \frac{dA}{dV}\frac{dV}{dM} = \frac{a(\vec{r},t)}{\rho(\vec{r},t)}\,.$$

Now we can derive from (2.8) a local balance equation for $\hat{a}$. To stay in the picture of small cells, which are locally in equilibrium (Figure 2.2), these cells have now all the same mass (e. g., one mole) but different volumes. With

$$a = \rho\,\hat{a} \tag{2.20}$$

equation (2.8) reads

$$\rho\frac{\partial\hat{a}}{\partial t} + \hat{a}\frac{\partial\rho}{\partial t} + \nabla\cdot\vec{j}_A = q_A\,.$$

On the l. h. s., we substitute $\partial\rho/\partial t$ by (2.11),

$$\text{l. h. s} = \rho\frac{\partial\hat{a}}{\partial t} - \hat{a}\nabla\cdot(\rho\,\vec{v}) + \nabla\cdot\vec{j}_A = \rho\frac{\partial\hat{a}}{\partial t} + \nabla\cdot(\vec{j}_A - \hat{a}\rho\,\vec{v}) + \rho\vec{v}\cdot\nabla\hat{a}\,.$$

Using the material derivative, we finally obtain a local balance equation for $\hat{a}$:

$$\boxed{\rho\frac{d\hat{a}}{dt} + \nabla\cdot\vec{a} = q_A}\,. \tag{2.21}$$

Here,

$$\vec{a} \equiv \vec{j}_A - \rho\hat{a}\vec{v} \tag{2.22}$$

defines the conductive flux density of A, denoted earlier in (2.10) with $\vec{j}_A^D$.

2.1.5 Specific balance equations

By the definition (2.20), the specific mass density is one. To find the specific equivalence of the continuity equation (2.11), we must use the specific volume $\hat{v} = 1/\rho$. Inserting into the balance equations derived earlier in Sections 2.1.3.1–4 for $\rho = 1/\hat{v}$, $\rho\vec{v} = \rho\hat{\vec{p}}$, $u = \rho\hat{u}$, and $s = \rho\hat{s}$, we obtain

$$\rho\frac{d}{dt}\left(\frac{1}{\rho}\right) = \nabla\cdot\vec{v}, \tag{2.23a}$$

$$\rho\frac{d\vec{v}}{dt} = \nabla\cdot\underline{\tau} + \vec{f} - \nabla p, \tag{2.23b}$$

$$\rho\frac{d\hat{u}}{dt} = -\nabla\cdot\vec{j}^Q + q_U, \tag{2.23c}$$

$$\rho\frac{d\hat{s}}{dt} = -\nabla\cdot\vec{s} + \sigma. \tag{2.23d}$$

In (2.23a), we use the notation $1/\rho$ instead of $\hat{v}$ for the specific volume, in (2.23b) $\vec{v}$ instead of $\hat{\vec{p}}$ for the specific momentum.

2.2 Entropy production of a one-component system

2.2.1 Irreversibility and entropy production

Gibbs fundamental relation (1.23) was derived for reversible processes. Using the second law in the form (1.21), we obtain the entropy change

$$dS = \frac{1}{T}dU + \frac{p}{T}dV + dS_{\text{irr}}\,. \tag{2.24}$$

For simplicity, we start with a simple system, e. g., a gas or a liquid of one component. To determine the irreversible non-negative part, the special processes must be known.

The strategy to compute dS_{irr} is the following: one assumes local equilibrium in each small cell at position $\vec{r}$ and time t. Then the variables in (1.23) become space and time dependent as explained above, but, even more important to note: equilibrium means that in each cell only (locally) reversible processes take place, and the reversible form (1.23) holds at each small cell at point $\vec{r}$. Under this assumption irreversibility originates only from diffusive fluxes (of momentum, heat) between the cells. Thus we are allowed to compute dS/dt from

$$\frac{dS}{dt} = \frac{1}{T}\frac{dU}{dt} + \frac{p}{T}\frac{dV}{dt} \tag{2.25}$$

and compare it with the local balance equation (2.23d) for the entropy derived in the previous section, where all fluxes, etc., have been included. Then the expression σ reflects the inner entropy sources, and integrating it over the whole domain corresponds to the rate of increase of S due to irreversible processes

$$\Sigma = \frac{\Delta S_{\text{irr}}}{\Delta t} = \int_V d^3\vec{r}\,\sigma\,.$$

Note that also stationary states do produce entropy if they are not in equilibrium. We mention a thermally conducting slab between two heat reservoirs with different temperatures.

2.2.2 One-component system

Starting point is Gibbs fundamental equation (1.23) or (2.25) for a one-component system, which we write now for *specific* variables ($\hat{v} = 1/\rho$),

$$\frac{d\hat{s}}{dt} = \frac{1}{T}\frac{d\hat{u}}{dt} + \frac{p}{T}\frac{d}{dt}\left(\frac{1}{\rho}\right). \tag{2.26}$$

From (2.23a) the last term is equal to

$$\frac{p}{\rho T}\nabla\cdot\vec{v}\,.$$

The total specific energy $\hat{e}$ of a subsystem moving with $\vec{v}$ is its specific internal (thermodynamic) energy plus its specific kinetic energy:

$$\hat{e} = \hat{u} + \frac{v^2}{2} \tag{2.27}$$

and

$$\frac{d\hat{u}}{dt} = \frac{d\hat{e}}{dt} - \frac{d}{dt}\left(\frac{v^2}{2}\right). \tag{2.28}$$

On the other hand, the balance equation for the total energy reads

$$\rho\frac{d\hat{e}}{dt} + \nabla\cdot\vec{e} = \vec{v}\cdot\vec{f}\,, \tag{2.29}$$

where the source term q_E is equal to the work done by external forces per time unit. The kinetic energy changes are found by multiplying (2.23b) with $\vec{v}$:

$$\rho \frac{d}{dt}\left(\frac{v^2}{2}\right) = \vec{v} \cdot \nabla \cdot \underline{\tau} + \vec{v} \cdot \vec{f} - \vec{v} \cdot \nabla p \,. \tag{2.30}$$

On the r. h. s., we write

$$\vec{v} \cdot \nabla \cdot \underline{\tau} = \nabla \cdot (\vec{v} \cdot \underline{\tau}) - \underline{\tau} : \nabla \vec{v}, \tag{2.31}$$

where ":" denotes the double product of two tensors defined as

$$\underline{A} : \underline{B} = \sum_{ij} A_{ij} B_{ij}, \tag{2.32}$$

and the components of the tensor $\nabla \vec{v}$ read

$$(\nabla \vec{v})_{ij} = \frac{\partial v_j}{\partial x_i} \,.$$

If $\underline{\tau}$ is symmetric (justification follows below), we may substitute

$$\underline{\tau} : \nabla \vec{v} = \frac{1}{2} \underline{\tau} : (\nabla \vec{v} + (\nabla \vec{v})^T) \,,$$

$(\nabla \vec{v})^T$ being the transpose of $(\nabla \vec{v})^T$. With the shear tensor $\underline{\epsilon}$ defined as

$$\epsilon_{ij} = \frac{1}{2}\left(\frac{\partial v_i}{\partial x_j} + \frac{\partial v_j}{\partial x_i}\right), \tag{2.33}$$

we may finally write

$$\underline{\tau} : \nabla \vec{v} = \underline{\tau} : \underline{\epsilon} \,.$$

Inserting (2.29) and (2.30) into (2.28) yields

$$\begin{aligned} \rho \frac{d\hat{u}}{dt} &= \underline{\tau} : \underline{\epsilon} + \vec{v} \cdot \nabla p - \nabla \cdot \vec{e} - \nabla \cdot (\vec{v} \cdot \underline{\tau}) \\ &= \underline{\tau} : \underline{\epsilon} + \nabla \cdot (p \vec{v}) - p \nabla \cdot \vec{v} - \nabla \cdot \vec{e} - \nabla \cdot (\vec{v} \cdot \underline{\tau}) \\ &= -\nabla \cdot (\vec{e} + \vec{v} \cdot \underline{\tau} - p \vec{v}) + \underline{\tau} : \underline{\epsilon} - p \nabla \cdot \vec{v}, \end{aligned} \tag{2.34}$$

which must be compared with (2.23c). From there, we identify the heat flux

$$\vec{j}^Q \equiv \vec{e} + \vec{v} \cdot \underline{\tau} - p \, \vec{v} \tag{2.35}$$

and the heat sources (due to friction and volume deformation)

$$q_U = \underline{\tau} : \underline{\epsilon} - p\,\nabla \cdot \vec{v}\,.$$

Next we insert (2.34) into (2.26) and obtain the specific entropy change as

$$\rho \frac{d\hat{s}}{dt} = \frac{1}{T}\left(\underline{\tau} : \underline{\epsilon} - \nabla \cdot \vec{j}^Q\right). \tag{2.36}$$

To cast (2.36) into the form of (2.23d), we write it as

$$\rho \frac{d\hat{s}}{dt} = -\nabla \cdot \left(\frac{\vec{j}^Q}{T}\right) + \vec{j}^Q \cdot \nabla\left(\frac{1}{T}\right) + \frac{\underline{\tau} : \underline{\epsilon}}{T}\,. \tag{2.37}$$

Comparison with (2.23d) finally yields

$$\boxed{\sigma = \vec{j}^Q \cdot \nabla\left(\frac{1}{T}\right) + \frac{\underline{\tau} : \underline{\epsilon}}{T}} \tag{2.38}$$

for the entropy production and

$$\vec{s} = \frac{\vec{j}^Q}{T} \tag{2.39}$$

for the conductive entropy flux.

2.3 Entropy production of a multi-component system

2.3.1 Definitions

Now let us consider a mixture of k components, where chemical reactions between the components shall be possible. Assume again local equilibrium in a small cell (volume ΔV), but now with different species inside. For simplicity, we begin with $k = 2$, Figure 2.6. Let Δn_i be the particle (or mole) number, M_i the particle (or molar) mass, and $\vec{v}_i$ the mean velocity of the particles of component i in ΔV. With the partial mass densities $\rho_i = M_i \Delta n_i / \Delta V$, we find for the total mass $\Delta M = \Delta n_1 M_1 + \Delta n_2 M_2$ and

$$\rho = \Delta M / \Delta V = \rho_1 + \rho_2 \tag{2.40}$$

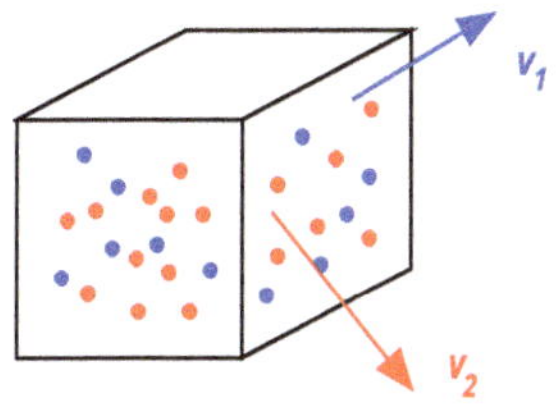

Figure 2.6: In a two component system, in each small volume ΔV two different kinds of particles exist. They have different mean velocities $\vec{v}_i$ and different particle densities $\Delta n_i M_i / \Delta V$.

for the mass density. The barycentric velocity is given as

$$\vec{v} = \frac{\Delta n_1 M_1 \vec{v}_1 + \Delta n_2 M_2 \vec{v}_2}{\Delta M} = \frac{\rho_1 \vec{v}_1 + \rho_2 \vec{v}_2}{\rho} . \tag{2.41}$$

The specific densities (or mass fractions) are defined as

$$c_i \equiv \hat{\rho}_i = \frac{\Delta m_i}{\Delta M} = \frac{\rho_i}{\rho} \tag{2.42}$$

with the mass $\Delta m_i = M_i \Delta n_i$ of component i in ΔV. With (2.40), it follows that

$$c_1 + c_2 = 1 . \tag{2.43}$$

Sometimes it is more convenient to use the mole fractions (as in Section 1.5.8),

$$X_i \equiv \frac{\Delta n_i}{\Delta n} \tag{2.44}$$

with the total mole number $\Delta n = \sum_i \Delta n_i$, or the molar densities

$$\tilde{c}_i \equiv \frac{\Delta n_i}{\Delta V} . \tag{2.45}$$

With the relation (2.22) and $\hat{a} = \hat{\rho}_i, \vec{j}_A = \rho_i \vec{v}_i$, the conductive (or better "diffusive") mass flux reads

$$\vec{J}_i = \rho_i (\vec{v}_i - \vec{v}),$$

and with (2.41)

$$\vec{J}_1 + \vec{J}_2 = 0 . \tag{2.46}$$

The local balance equations (2.21) for the specific densities c_i read

$$\rho \frac{dc_j}{dt} + \nabla \cdot \vec{J}_j = q_j , \tag{2.47}$$

where the sources and sinks q_j are due to chemical reactions. Summing up (2.47) over j yields with (2.43) and (2.46)

$$\sum_j^k q_j = 0 , \tag{2.48}$$

which expresses mass conservation for each chemical reaction.

External forces may act in different ways on the different components. Take for instance gravity, then

$$\vec{f}_i = \rho_i \vec{g}$$

is the force density on component i and $\vec{g}$ the gravitational acceleration. Clearly,

$$\vec{f} = \vec{f}_1 + \vec{f}_2$$

is the total force density.

We summarize the important variable for a multi-component system:

ρ_j	density [kg/m^3]	$\sum_j \rho_j = \rho$	total density
$\vec{v}_j$	velocity [m/s]	$\frac{1}{\rho}\sum_j \rho_j \vec{v}_j = \vec{v}$	barycentric velocity
$\vec{J}_j = \rho_j(\vec{v}_j - \vec{v})$	diffusive mass flux [kg/(m^{2}s)]	$\sum_j \vec{J}_j = 0$	
$c_j = \rho_j/\rho$	rel. concentration, spec. density, mass fraction	$\sum_j c_j = 1$	
$\tilde{c}_j = \rho c_j/M_j$	molar concentration [1/m^3]		
$X_j = n_j/n$	mole fraction	$\sum_j X_j = 1$	
$\hat{\mu}_j = \mu_j/M_j$	spec. chem. potential [J/(kg mol)]		M_j: molar mass
f_j	external force [N/m^3]	$\sum_j \vec{f}_j = \vec{f}$	total force

All the above quantities are defined in each small "equilibrium cell" or subsystem and depend in general on space and time. In addition to conductive heat and momentum fluxes, we have now also a diffusive (conductive) mass flux for each component.

2.3.2 Gibbs fundamental equation

With (1.60), equation (2.26) takes the extended form

$$\rho \frac{d\hat{s}}{dt} = \frac{\rho}{T}\frac{d\hat{u}}{dt} + \frac{\rho p}{T}\frac{d}{dt}\left(\frac{1}{\rho}\right) - \frac{\rho}{T}\sum_j^k \hat{\mu}_j \frac{dc_j}{dt}, \tag{2.49}$$

and (2.29) reads

$$\rho \frac{d\hat{e}}{dt} + \nabla \cdot \vec{e} = \sum_j^k \vec{v}_j \cdot \vec{f}_j \,. \tag{2.50}$$

Therefore, instead of (2.34), we receive

$$\rho \frac{d\hat{u}}{dt} = -\nabla \cdot (\vec{e} + \vec{v} \cdot \underline{\tau} - p\vec{v}) + \underline{\tau} : \underline{\epsilon} - p\nabla \cdot \vec{v} + \sum_j^k \vec{v}_j \cdot \vec{f}_j - \vec{v} \cdot \vec{f} \,. \tag{2.51}$$

This yields extra terms on the r. h. s. of (2.49) of the form

$$\frac{1}{T}\sum_j^k(\vec{v}_j\cdot\vec{f}_j-\vec{v}\cdot\vec{f}_j)=\frac{1}{T}\sum_j^k\frac{\vec{J}_j\cdot\vec{f}_j}{\rho_j} \tag{2.52}$$

with the diffusive mass flux $\vec{J}_j$ of component j.

Finally, we compute the contributions of the last expression in (2.49) using the balance equations (2.47):

$$\frac{\rho}{T}\sum_j^k\hat{\mu}_j\frac{dc_j}{dt}=\frac{1}{T}\sum_j^k\hat{\mu}_j(-\nabla\vec{J}_j+q_j)\,. \tag{2.53}$$

2.3.3 Chemical reactions

To compute the source terms q_j on the right hand side of (2.53), chemical reactions must be taken into account. As an example, consider a certain reaction, namely the production of oxyhydrogen gas ("knallgas") via

$$2\ \mathrm{H}_2+\mathrm{O}_2\to 2\ \mathrm{H}_2\mathrm{O}\,. \tag{2.54}$$

From there we find the stoichiometric coefficients

$$\nu_{\mathrm{H}_2}=-2,\quad \nu_{\mathrm{O}_2}=-1,\quad \nu_{\mathrm{H}_2}\mathrm{O}=2\,,$$

where $\nu_j<0$ accounts for the annihilation (left hand side), $\nu_j>0$ for the production of $|\nu_j|$ molecules of component j (see also Section 1.5.8). Now taking the balance equation for c_{H_2}, a source (or better sink) term of the form

$$q_{\mathrm{H}_2}=\omega\ \nu_{\mathrm{H}_2}M_{\mathrm{H}_2}<0$$

results, where M_{H_2} is the (constant) molar mass of H_2 and ω the velocity of the reaction (density, space and time dependent). The general form of the source terms thus reads

$$q_j=\omega\ \nu_jM_j,$$

where

$$\omega=\omega(c_1\ldots c_k)$$

may depend in a nonlinear way on the concentrations; see the considerations of Section 2.4.6 below. For the special reaction (2.54), one may assume

$$\omega\sim(c_{\mathrm{H}_2})^2\cdot c_{\mathrm{O}_2},$$

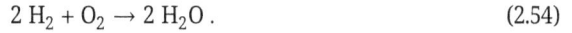

because two molecules of H_2 and one molecule of O_2 are needed at the same location. Normally one has a system comprised of a certain number, say R, of reactions, taking place with different velocities, and

$$\nu_j \to \nu_{rj}, \quad \omega \to \omega_r .$$

Then the sources read

$$q_j = \sum_r^R \omega_r \, \nu_{rj} M_j . \tag{2.55}$$

Finally, taking (2.48) and inserting (2.55) yields

$$\sum_j^k \nu_{rj} M_j = 0 ,$$

which reflects mass conservation for each reaction r.

2.3.4 Entropy production

Inserting (2.55) in (2.53), we find the additional entropy production rates, absent in a one-component system

$$\nabla \cdot \left(\frac{1}{T} \sum_j^k \hat{\mu}_j \vec{J}_j \right) - \sum_{j=1}^k \vec{J}_j \cdot \nabla \left(\frac{\hat{\mu}_j}{T} \right) - \frac{1}{T} \sum_{r=1}^R \omega_r \sum_j^k \nu_{rj} \hat{\mu}_j M_j . \tag{2.56}$$

The first term is conservative and can be absorbed into the conductive entropy flux (2.39)

$$\vec{s} = \frac{\vec{j}^Q}{T} - \frac{1}{T} \sum_j^k \hat{\mu}_j \vec{J}_j . \tag{2.57}$$

The entropy production rate for a multi-component system with chemical reactions reads with (2.38), (2.52), and (2.56)

$$\boxed{\sigma = \vec{j}^Q \cdot \nabla \left(\frac{1}{T} \right) + \frac{\underline{\tau} : \underline{\epsilon}}{T} + \sum_{j=1}^k \vec{J}_j \cdot \left(\frac{\vec{f}_j}{\rho_j T} - \nabla \left(\frac{\hat{\mu}_j}{T} \right) \right) - \sum_{r=1}^R \omega_r \frac{A_r}{T}} , \tag{2.58}$$

where the *affinities* are defined as

$$A_r \equiv \sum_j^k \nu_{rj} \hat{\mu}_j M_j = \sum_j^k \nu_{rj} \mu_j . \tag{2.59}$$

As long as $A_r < 0$, the reaction r runs from left to right ($\omega_r > 0$); for $A_r > 0$ it goes backwards. In both cases, the product $\omega_r A_r$ is negative, and entropy is produced as long as the reaction runs. From the considerations in Section 1.5.8, equation (1.66), we see that in thermal equilibrium $A_r = 0$ holds.

2.4 Onsager relations

2.4.1 Fluxes and forces

For simplicity, we start with a one-component system. Looking at the production rate (2.38), we see that it can be written as a sum of products,

$$
\sigma = \underbrace{\vec{j}^Q}_{\text{flux}} \cdot \underbrace{\nabla\left(\frac{1}{T}\right)}_{\text{force}} + \underbrace{\underline{\tau}}_{\text{flux}} : \underbrace{\frac{\underline{\epsilon}}{T}}_{\text{force}} .
\tag{2.60}
$$

The fluxes also occur in the conservation equations for internal energy and momentum, respectively. From (2.60) we can identify the forces, but the fluxes are still unknown. First we write (2.60) in the general form that holds also for (2.58):

$$
\sigma = \sum_i J_i \bullet X_i ,
\tag{2.61}
$$

where J_i stands for $\vec{j}^Q$, $\underline{\tau}$, $\vec{J}_i$ or ω_r, and X_i is the corresponding "force." The product $\bullet$ depends on the character of the variables J_i, X_i. For scalars, it is a simple product, for vectors a scalar product, for tensors a double product (2.32), and so on.

Phenomenologically, a flux needs a driving force. A heat flux exists if a temperature gradient exists; a momentum flux is related to shear stresses in a fluid, a mass flux to a concentration gradient, and the reaction rates are linked to the affinities. If the forces vanish, so do the fluxes, and it is straightforward to expand the fluxes with respect to the forces. The linear branch of non-equilibrium thermodynamics considers only the first terms (the linear ones) of this expansion, according to

$$
J_i = \sum_j L_{ij} X_j,
\tag{2.62}
$$

where the matrix $\underline{L}$ comprises the phenomenological or Onsager[1] coefficients.

In 1931, Onsager showed [4] that the coefficients must have the symmetry

$$
L_{ij} = \alpha_i \alpha_j L_{ji}, \quad \alpha_i = \begin{cases} 1 & \text{if } X_i(t) = X_i(-t), \\ -1 & \text{if } X_i(t) = -X_i(-t). \end{cases}
\tag{2.63}
$$

---

[1] Lars Onsager, Norwegian physicist, 1903–1976.

Finally, inserting (2.62) in (2.61) yields for the entropy production

$$\sigma = \sum_{ij} L_{ij} X_i \, X_j \, . \tag{2.64}$$

Since $\sigma \geq 0$, the coefficient matrix $\underline{L}$ must be positive semi-definite to fulfill the second law of thermodynamics.

Note that (2.62) is much more general than simply assuming

$$J_i = L_i X_i,$$

because it relates all forces to all fluxes. Thus, not only a temperature gradient but also a concentration gradient in a mixture may contribute to the energy flux; these effects are called "cross effects." We mention thermodiffusion, the Soret effect, where a temperature gradient causes a mass flux, and the cross-related Dufour effect, where a concentration gradient creates a temperature gradient. We shall discuss both issues in more detail later.

2.4.2 Curie's principle

Under the assumption that the entropy production (2.61) is a scalar with respect to rotation and translation of the coordinate system, only quantities that have the same behavior under spatial transformations can couple to each other, say scalar to scalar, vector to vector, tensor to tensor. This is Curie's[2] principle; it reduces the number of coefficients drastically.

2.4.3 Onsager coefficients

Now we are in the position to relate the fluxes to the forces. According to Curie's principle, we have to distinguish between scalars, vectors, and tensors.

Scalars: in (2.60) only the traces of $\underline{\tau}$ and $\underline{\epsilon}$ transform like a scalar. Thus we put

$$\mathrm{Tr}\,\underline{\tau} = L_s \frac{\mathrm{Tr}\,\underline{\epsilon}}{T} \tag{2.65}$$

with the first Onsager coefficient L_s.

Vectors: the energy flux is related to the force $\nabla(1/T)$ with

2 Pierre Curie, French physicist, 1859–1906.

$$\vec{j}^Q = L_v \nabla\left(\frac{1}{T}\right) = -\frac{L_v}{T^2}\nabla T \,. \tag{2.66}$$

Identifying the Onsager coefficient

$$L_v = \lambda T^2 \,, \tag{2.67}$$

we immediately receive Fourier's law with λ being the thermal conductivity:

$$\vec{j}^Q = -\lambda \nabla T \,. \tag{2.68}$$

Tensors: the momentum flux $\underline{\tau}$ is related to the force $\underline{\epsilon}/T$. There are two possible tensors that are linear in $\underline{\epsilon}$. One is $\underline{\epsilon}$ itself, the other is $\underline{1}\,\mathrm{Tr}\,\underline{\epsilon}$, with $\underline{1}$ as the identity matrix. We write the force as a superposition

$$\underline{\tilde{\epsilon}} = \underline{\epsilon} + \alpha\,\underline{1}\,\mathrm{Tr}\,\underline{\epsilon}$$

and relate the momentum flux to

$$\underline{\tau} = \frac{L_t}{T}\,\underline{\tilde{\epsilon}} \,.$$

Inserting this in (2.65) yields

$$\alpha = \frac{L_s - L_t}{3L_t} \,,$$

and finally

$$\underline{\tau} = \frac{L_t}{T}\left(\underline{\epsilon} - \frac{1}{3}\underline{1}\,\nabla\cdot\vec{v}\right) + \frac{1}{3}\underline{1}\,\frac{L_s}{T}\,\nabla\cdot\vec{v} \,, \tag{2.69}$$

where we substituted

$$\mathrm{Tr}\,\underline{\epsilon} = \nabla\cdot\vec{v}$$

from the definition of the shear tensor (2.33). For a Newtonian fluid (see textbooks on hydrodynamics, e. g., [8])

$$\underline{\tau} = 2\eta\,\underline{\epsilon} + \underline{1}\left(\zeta - \frac{2}{3}\eta\right)\nabla\cdot\vec{v} \,. \tag{2.70}$$

Here, η is the dynamic viscosity and ζ the second viscosity. If we identify the Onsager coefficients with

$$L_t = 2T\eta, \quad L_s = 3T\zeta \,,$$

(2.69) has the form of (2.70).

Introducing the trace-free shear tensor

$$\underline{\epsilon}' = \underline{\epsilon} - \frac{1}{3}\underline{1}\,\mathrm{Tr}\,\underline{\epsilon}, \tag{2.71}$$

one can write the stress tensor also in the form of

$$\underline{\tau} = 2\eta\,\underline{\epsilon}' + \zeta\,\underline{1}\,\nabla\cdot\vec{v}\,. \tag{2.72}$$

Of course the Onsager coefficients are introduced solely on phenomenological grounds, and the *transport coefficients* λ, η, ζ cannot be computed without further knowledge of the materials. Here one needs more fundamental theories and models or experimental data. In part II of the book, we shall present a derivation of the transport coefficients for a perfect gas based on statistical mechanics.

2.4.4 Entropy production and second law

Inserting (2.72) and (2.66) into (2.60), we finally obtain a closed expression for the entropy production rate

$$\sigma = \lambda T^2\left(\nabla\left(\frac{1}{T}\right)\right)^2 + \frac{1}{T}\left(2\eta\,\underline{\epsilon}' : \underline{\epsilon} + \zeta\,(\nabla\cdot\vec{v})^2\right), \tag{2.73}$$

where we used $\underline{\epsilon} : \underline{1} = \mathrm{Tr}\,\underline{\epsilon} = \nabla\cdot\vec{v}$. The 2nd law requires that σ cannot be negative. This is clearly true for the first term of (2.73). If we use in the second term again the decomposition (2.71), we find

$$2\eta\,\underline{\epsilon}' : \underline{\epsilon} + \zeta\,(\nabla\cdot\vec{v})^2 = 2\eta\left(\underline{\epsilon}' : \underline{\epsilon}' + \frac{1}{3}\,\mathrm{Tr}\,\underline{\epsilon}\,(\underline{\epsilon}' : \underline{1})\right) + \zeta\,(\nabla\cdot\vec{v})^2 = 2\eta\,\underline{\epsilon}' : \underline{\epsilon}' + \zeta\,(\nabla\cdot\vec{v})^2 \geq 0,$$

where we used $\underline{\epsilon}' : \underline{1} = \mathrm{Tr}\,\underline{\epsilon}' = 0$.

2.4.5 Onsager coefficients for a multi-component system

For a multi-component system, additional forces are caused by chemical reactions and concentration gradients.

Scalars: in addition we have the forces $-A_r/T$, generalizing (2.65) to

$$\mathrm{Tr}\,\underline{\tau} = 3\,\zeta\;\mathrm{Tr}\,\underline{\epsilon} + \sum_r^R L_r\left(-\frac{A_r}{T}\right). \tag{2.74}$$

For the fluxes ω_r, we have

$$\omega_r = \sum_q^R L_{rq}\left(-\frac{A_q}{T}\right) + L_r' \frac{\mathrm{Tr}\,\underline{\epsilon}}{T} . \tag{2.75}$$

The coefficients must fulfill (2.63):

$$L_r = -L_r', \quad L_{rq} = L_{qr} .$$

The minus is due to $\underline{\epsilon} = -\underline{\epsilon}(-t)$ but $A(t) = A(-t)$.

Vectors: due to additional forces, the heat flux (2.66) gives rise to

$$\vec{j}^Q = \lambda T^2 \nabla\left(\frac{1}{T}\right) + \sum_{j=1}^k \ell_j \left(\frac{\vec{f}_j}{\rho_j T} - \nabla\left(\frac{\hat{\mu}_j}{T}\right)\right) . \tag{2.76}$$

The mass fluxes read

$$\vec{J}_i = \ell_i' \, \nabla\left(\frac{1}{T}\right) + \sum_{j=1}^k \ell_{ij} \left(\frac{\vec{f}_j}{\rho_j T} - \nabla\left(\frac{\hat{\mu}_j}{T}\right)\right) , \tag{2.77}$$

with the symmetries

$$\ell_j = \ell_j', \quad \ell_{ij} = \ell_{ji} , \tag{2.78}$$

and

$$\ell_{ii} > 0 .$$

The ℓ_{ii} link the concentration gradient of i to the flux of i; the ℓ_{ij}, $i \neq j$ describe cross-diffusion. The cross terms ℓ_j account for a heat flux due to concentration gradients, (2.76), and a mass flux due to a temperature gradient, equation (2.77). Both effects belong to the subject of thermodiffusion; the first one is called *Dufour effect*, the latter *Soret effect*.

Tensors: finally due to (2.74), the material law (2.70) has to be extended to

$$\underline{\tau} = \underline{1}\left(\zeta - \frac{2}{3}\eta\right)\nabla \cdot \vec{v} + 2\eta\, \underline{\epsilon} - \underline{1} \sum_r^R L_r \frac{A_r}{T} . \tag{2.79}$$

2.4.6 A note on reaction velocity

We repeat (2.75) and neglect for the moment the term with $\mathrm{Tr}\,\underline{\epsilon}$:

$$\omega_r = -\frac{1}{T} \sum_q^R L_{rq} A_q . \tag{2.80}$$

According to Onsager, the fluxes ω_r are linearly related to the forces A_q, and thus linear in μ_j. In Section 2.3.3 however, we assumed for ω a nonlinear ansatz of the form

$$\omega_r \sim \prod_{i=1}^{\ell} c_i^{|\nu_{ri}|} = K_r \prod_{i=1}^{\ell} \tilde{c}_i^{|\nu_{ri}|}, \tag{2.81}$$

where $c_1 \dots c_\ell$ denote the concentrations of the reactants on the left hand side of reaction r, $\tilde{c}_1 \dots \tilde{c}_\ell$ their molar densities (2.45) (already introduced in Section 1.5.8) and K_r the reaction rate.

This ansatz is the classical one coming from reaction kinetics necessary to obtain nonlinear reaction diffusion equations and to describe the temporal evolution of non-equilibrium reactions.

How are (2.80) and (2.81) related? Consider the general form of a reaction

$$\sum_{i=1}^{\ell_r} (-\nu_{ri}) \mathrm{B}_i \underset{K_r'}{\overset{K_r}{\rightleftarrows}} \sum_{i=\ell_r+1}^{k_r} \nu_{ri} \mathrm{B}_i \tag{2.82}$$

with $\mathrm{B}_1 \dots \mathrm{B}_{\ell_r}$ as the reactants and $\mathrm{B}_{\ell_r+1} \dots \mathrm{B}_{k_r}$ as the products. Then from (2.81), we find

$$\omega_r = K_r \prod_{i=1}^{\ell_r} \tilde{c}_i^{|\nu_{ri}|} - K_r' \prod_{i=\ell_r+1}^{k_r} \tilde{c}_i^{\nu_{ri}}, \tag{2.83}$$

where the second product accounts for the backwards reaction. In equilibrium, $\omega_r = 0$, and thus

$$\frac{\prod_{i=\ell_r+1}^{k_r} \tilde{c}_i^{\nu_{ri}}}{\prod_{i=1}^{\ell_r} \tilde{c}_i^{|\nu_{ri}|}} = \tilde{K}_r$$

with $\tilde{K}_r = \frac{K_r}{K_r'}$. This has the form of the law of mass action (1.67), but now formulated in terms of the molar densities, leading to a different constant $\tilde{K}$. Vanishing ω_r means *detailed balance*. This is the case when forward and backward reaction in each reaction step r are in balance.

To see the relation between (2.80) and (2.83), we write (2.83) as

$$\omega_r = \omega_r^{(+)} \left(1 - \frac{1}{\tilde{K}_r} \prod_{i=1}^{k_r} \tilde{c}_i^{\nu_{ri}} \right) \tag{2.84}$$

with

$$\omega_r^{(+)} = K_r \prod_{i=1}^{\ell_r} \tilde{c}_i^{-\nu_{ri}}$$

as velocity of the forward reaction.

To establish the link between A_r and ω_r, we assume a nonlinear relation of the form

$$\omega_r = \omega_r^0 \left(1 - e^{A_r/RT}\right) \tag{2.85}$$

which will be justified a posteriori in providing the correct kinetic reaction velocities and the correct linearization close to equilibrium. Checking the entropy production rate

$$\sigma(A_r) = -\omega_r \frac{A_r}{T} = -\omega_r^0 \frac{A_r}{T}(1 - e^{A_r/RT}),$$

one can easily show that for $\omega_r^0 > 0$ the only minimum of σ is at equilibrium, $A_r = 0$, $\sigma = 0$. Since $\sigma \to \infty$ for $A_r \to \pm\infty$, the ansatz (2.85) is in agreement with the second law.

For the evaluation of A_r from (2.59), we take the chemical potentials of an ideal gas mixture (1.59) and formulate them first with the molar densities. Therefore, we write

$$\begin{aligned} \mu_j(p, T, n_i) &= c_p^{(j)} T - c_p^{(j)} T \ln\left(\frac{T}{T_0}\right) - Ts_0 + RT \ln\left(\frac{p}{p_0}\right) + RT \ln\left(\frac{n_j}{n}\right) \\ &= \tilde{g}_j(T) + RT \ln\left(\frac{pT_0}{p_0 T} \frac{\Delta n_j}{\Delta n}\right) \end{aligned} \tag{2.86}$$

with

$$\tilde{g}_j(T) = c_p^{(j)} T - c_v^{(j)} T \ln\left(\frac{T}{T_0}\right) - Ts_0$$

as a function of temperature only. Here, $\Delta \ldots$ denotes the values in the small cell ΔV, according to the concept of local equilibrium. From the ideal gas equation, we have

$$\frac{p\Delta V}{p_0 \Delta V_0} = \frac{\Delta n}{\Delta n_0} \frac{T}{T_0},$$

and we can convert the last term of (2.86) into

$$\ln\left(\frac{pT_0}{P_0 T} \frac{\Delta n_j}{\Delta n}\right) = \ln\left(\frac{\Delta V_0}{\Delta V} \frac{\Delta n_j}{\Delta n_0}\right),$$

from where

$$\mu_j(p, T, \tilde{c}_i) = \tilde{g}_j(T) + RT \ln\left(\frac{\tilde{c}_j}{\tilde{c}}\right). \tag{2.87}$$

Here, $\tilde{c} = \Delta n_0 / \Delta V_0$ denotes some constant reference molar density.

From (2.59), we obtain

$$A_r = \sum_j^{k_r} \nu_{rj} \mu_j = \sum_j^{k_r} \nu_{rj} \tilde{g}_j(T) + RT \sum_j^{k_r} \nu_{rj} \ln\left(\frac{\tilde{c}_j}{\tilde{c}}\right). \tag{2.88}$$

Inserting this into (2.85) finally yields

$$\omega_r = \omega_r^0 \left(1 - a_r \prod_i^{k_r} \tilde{c}_i^{\nu_{ri}} \right) \tag{2.89}$$

with the abbreviation

$$a_r = \tilde{c}^{-\sum_j \nu_{rj}} \exp\left(\sum_i^{k_r} \nu_{ri} \tilde{g}_i / RT \right).$$

Comparing (2.89) with (2.84), we may easily identify

$$\omega_r^0 = \omega_r^{(+)}, \quad a_r = \frac{1}{\tilde{K}_r}.$$

In this way, the relation (2.85) yields the correct kinetic expressions for the reaction velocities. At equilibrium, $A_r = 0$, and close to equilibrium, we may expand (2.85) up to linear order

$$\omega_r = -\omega_r^{(+)} \frac{A_r}{RT} + O(A_r^2), \tag{2.90}$$

which turns out to be a special form of (2.80) with a diagonal Onsager coefficient matrix

$$L_{rq} = \frac{\omega_r^{(+)}}{R} \delta_{rq}.$$

As a conclusion, we see that we need a relation between fluxes and forces that go over the linear ones (2.90) to describe the correct velocities used in reaction kinetics.

We finally note that constant concentrations do not always mean equilibrium. If a cyclic reaction system of the (symbolic) form

$$\begin{aligned} \mathrm{A} &\xrightarrow{K_1} \mathrm{B}, \\ \mathrm{B} &\xrightarrow{K_2} \mathrm{C}, \\ \mathrm{C} &\xrightarrow{K_3} \mathrm{A} \end{aligned} \tag{2.91}$$

takes place, the concentrations are constant if

$$\omega_1 = \omega_2 = \omega_3$$

is fulfilled. A steady solution would be

$$c_i = \alpha / K_i$$

with some positive α that can for instance be determined from $\sum_i c_i = 1$. In this case, the stationary solution is not in detailed balance, and the law of mass action cannot be used.

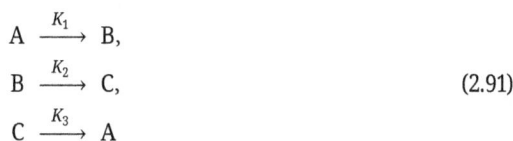

Since $\omega_r \neq 0$ and $A_r \neq 0$ (all reactions are supposed only running forward), this reaction scheme produces permanent entropy. Of course, this is only possible in an open system, where the reaction is feeded from outside.

A similar situation is given by a constantly feeded reaction system,

$$\begin{aligned} \mathrm{A} &\xrightarrow{K_1} \mathrm{B}, \\ \mathrm{B} &\xrightarrow{K_2} \mathrm{C}, \\ \mathrm{C} &\xrightarrow{K_3} \mathrm{D}, \end{aligned} \tag{2.92}$$

where A is kept constant and D is removed; a stationary solution is given as

$$c_2 = \frac{K_1}{K_2} c_1, \quad c_3 = \frac{K_1}{K_3} c_1 .$$

Also here, ω_r, $A_r \neq 0$ and $\sigma > 0$ for all time.

2.5 Problem: chemical reactions

Consider the following auto-catalytic reaction:

$$\mathrm{A} + \mathrm{B} \underset{K'}{\overset{K}{\rightleftharpoons}} 2\,\mathrm{A} . \tag{1}$$

The mass fractions of species A, B are c_1, c_2 and

$$c_1 + c_2 = 1 . \tag{2}$$

For the whole problem, spatial dependencies shall be neglected and the concentrations depend only on time (stirred reaction). The medium shall be at rest ($\vec{v} = 0$).

a) Determine the reaction velocity ω as a function of c_i for (1).
   Determine the stoichiometric coefficients ν_i. Neglect diffusive fluxes and formulate the rate equations for c_i, according to

$$\rho \dot{c}_i = q_i = \omega(c_1, c_2)\nu_i M_i, \quad i = 1, 2, \tag{3}$$

   where the dot denotes the time derivative, and the density ρ is constant (why?). With (2), find a relation for the molar masses M_A, M_B.

b) From (3), eliminate c_2 with (2). You should obtain an equation of the form

$$\dot{c} = a_1 c - a_2 c^2 \tag{4}$$

   with $c = c_1$. What do you obtain for a_i?

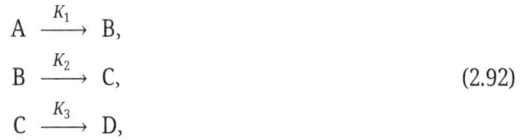

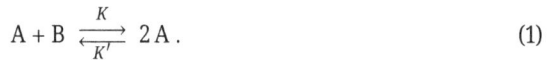

c) From (4), determine the equilibrium solutions $c = c_{\text{eq}} = \text{const}$. Can you say something on the stability of c_{eq}?

d) Integrate (4) (separation of variables) to find $c(t)$. What do you expect for $c(t \to \infty)$?

3 Transport equations

The last task left is to write down equations for the spatial and temporal evolution of the state variables

$$\hat{u}(\vec{r},t), \quad \rho(\vec{r},t), \quad \vec{v}(\vec{r},t), \quad c_j(\vec{r},t)\,.$$

From the equations of state, we can then compute temperature, pressure, and chemical potentials. The desired evolution equations are based on first principles: conservation for mass, momentum, and energy. They have been stated already in Chapter 2. With the linear Onsager relations for the fluxes postulated in Section 2.4, they can be closed. We start with the one-component system.

3.1 One-component system

3.1.1 Mass conservation

This is the continuity equation for the density (2.23a),

$$\frac{1}{\rho}\frac{d\rho}{dt} + \nabla\cdot\vec{v} = 0\,. \tag{3.1}$$

For an incompressible fluid, the continuity equation reduces to

$$\nabla\cdot\vec{v} = 0\,. \tag{3.2}$$

The mass flux has no diffusive part, and an Onsager relation is not needed.

3.1.2 Momentum conservation

Taking (2.23b) and substituting the diffusive flux $\underline{\tau}$ by the closure (2.70), we obtain

$$\rho\frac{d\vec{v}}{dt} = \eta\nabla^2\vec{v} + \left(\zeta + \frac{1}{3}\eta\right)\nabla(\nabla\cdot\vec{v}) + \vec{f} - \nabla p\,. \tag{3.3}$$

These are the *Navier–Stokes equations*. For an incompressible fluid, they simplify to

$$\rho\frac{d\vec{v}}{dt} = \eta\nabla^2\vec{v} + \vec{f} - \nabla p\,; \tag{3.4}$$

for an inviscid fluid $\eta = \zeta = 0$ and $\underline{\tau} = 0$, and we find

$$\rho\frac{d\vec{v}}{dt} = -\nabla p + \vec{f}\,, \tag{3.5}$$

the *Euler equations*.

https://doi.org/10.1515/9783110794137-004

For compressible fluids or gases, an equation of state of the form

$$p = p(\rho, T) \tag{3.6}$$

is needed.

Isothermal one-component fluids are completely described with (3.1), (3.3), and (3.6). If on the other hand the fluid is incompressible, this reduces to (3.2) and (3.4). Then (3.6) is no longer defined, and p must be computed from the Navier–Stokes equations. This can be achieved taking the divergence of (3.4) and making use of (3.2), leading to a Poisson equation,

$$\nabla^2 p = \nabla \cdot \vec{f} - \rho \, \nabla \cdot \left[(\vec{v} \cdot \nabla)\vec{v}\right] . \tag{3.7}$$

Appropriate boundary conditions for p are normally found from the Navier–Stokes equations evaluated at the boundaries; for more details see [5].

3.1.3 Energy conservation

Taking (2.23b) and closing with (2.68) and (2.70) yields

$$\rho \frac{d\hat{u}}{dt} = \lambda \nabla^2 T - p \, \nabla \cdot \vec{v} + 2\eta \, \underline{\epsilon} : \underline{\epsilon} + \left(\zeta - \frac{2}{3}\eta\right)(\nabla \cdot \vec{v})^2 . \tag{3.8}$$

In pure fluids or in solids, the inner energy can be approximated as a linear function of T and the caloric state equation reads

$$\hat{u} = c \, T$$

with the specific heat c. Assuming c as a constant, (3.8) turns into the heat equation, which allows computing a temperature distribution evolving in time. We state the heat equation here only for an incompressible fluid:

$$\frac{dT}{dt} = \kappa \nabla^2 T + \frac{2\eta}{\rho c} \, \underline{\epsilon} : \underline{\epsilon} , \tag{3.9}$$

where

$$\kappa = \frac{\lambda}{\rho c} \tag{3.10}$$

denotes the *thermal diffusivity* (not to confuse with the *thermal conductivity* λ).

Note that due to the material derivative, defined in (2.17), all the evolution equations stated above are nonlinear, opening the possibility for chaotic dynamics, self-organized pattern formation, and turbulence.

3.1.4 Application: entropy production of a slab

First we study the case of a solid where no material flow exists. Consider a slab with length L and cross section area A. Both ends of the slab are in contact with heat baths having different temperatures T_0, T_1; Figure 3.1. Let $T_1 > T_0$. A steady solution of (3.9) reads

$$T(z) = T_0 + \frac{\Delta T}{L} z, \quad \Delta T = T_1 - T_0 ,$$

where the side wall of the slab is assumed to be insulating. The steady heat flux along the slab is

$$j^Q = -\lambda \frac{\Delta T}{L} ,$$

and with (2.38)

$$\sigma = \frac{\lambda}{T^2(z)} \left(\frac{\Delta T}{L} \right)^2 .$$

Integrating over the volume of the slab yields

$$\Sigma = \frac{\lambda V}{T_0 T_1} \left(\frac{\Delta T}{L} \right)^2 \tag{3.11}$$

with volume $V = LA$.

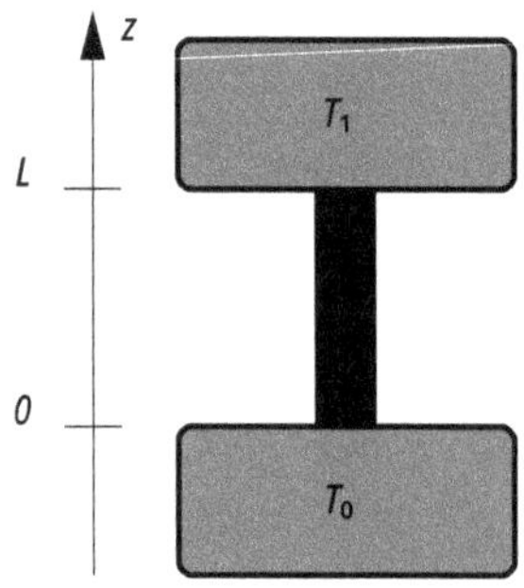

Figure 3.1: A solid slab connects two heat baths. If $T_1 > T_0$, heat flows irreversibly from the upper bath to the lower one, and entropy is produced constantly.

On the other hand, we can consider the heat amount per time taken from the upper bath and added to the lower one

$$\frac{\Delta Q}{\Delta t} = A \left| j^Q \right| .$$

This lowers the entropy of the upper bath by

$$\Delta S_1 = -\frac{\Delta Q}{T_1}$$

but increases the entropy of the lower one by

$$\Delta S_0 = \frac{\Delta Q}{T_0},$$

giving the total entropy change per time,

$$\frac{\Delta S_0 + \Delta S_1}{\Delta t} = \left(\frac{1}{T_0} - \frac{1}{T_1}\right) A \, |j^Q| = \frac{\lambda A}{T_0 T_1} \frac{\Delta T^2}{L},$$

which is the same as (3.11).

If we consider the whole system of slab and heat baths as a closed one, the upper heat bath loses less entropy than the lower one gains. According to the second law, the difference is produced by the non-equilibrium process of maintaining the heat flux inside the slab.

3.1.5 Application: entropy production of a pipe flow

Next we wish to compute the entropy production of a stationary, viscous, and incompressible flow through a cylindrical pipe, Figure 3.2.

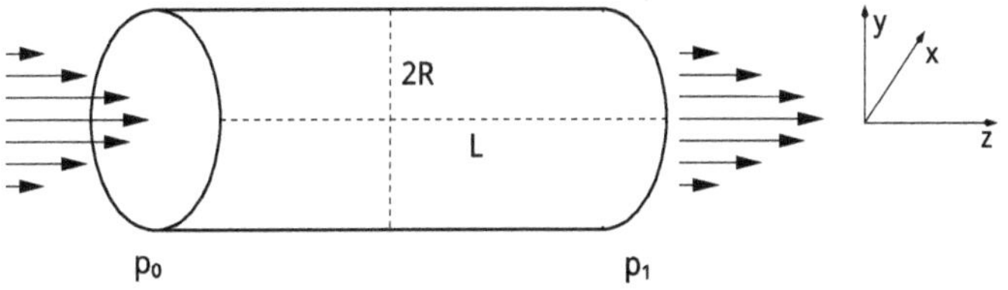

Figure 3.2: A stationary flow through a pipe with length L and radius R is driven by the pressure difference $\Delta p = p_0 - p_1$. Due to viscous friction, entropy is produced continuously.

(a) Isothermal case. If fluid and pipe are good heat conductors and in thermal contact with a heat bath of temperature T_0, the whole system can be approximately considered being isothermal, and from (2.38) only the "mechanical part"

$$\sigma = \frac{1}{T} (\underline{\tau} : \underline{\epsilon}) \tag{3.12}$$

survives, where $T = T_0$ is the temperature of the heat bath. For an incompressible fluid, the relation (2.70) leads to $\underline{\tau} = 2\eta\underline{\epsilon}$ and

$$\sigma = \frac{2\eta \, \underline{\epsilon} : \underline{\epsilon}}{T_0}. \tag{3.13}$$

Taking cylindrical coordinates, the stationary velocity field has the form

$$\vec{v} = v(r)\,\hat{e}_z$$

with $r = \sqrt{x^2 + y^2}$. On a rigid wall, the velocity must vanish, and

$$v(R) = 0\,.$$

The velocity is computed from (3.4) in cylindrical coordinates; see Appendix C.5.1.1

$$\eta \frac{1}{r}\frac{d}{dr}\left(r\frac{d}{dr}v(r)\right) = \frac{dp}{dz} = -\frac{\Delta p}{L}$$

as *Hagen–Poiseuille flow*

$$v(r) = \frac{\Delta p}{4L\eta}(R^2 - r^2)\,.$$

Then the only non-vanishing components of $\underline{\epsilon}$ are

$$\epsilon_{rz} = \epsilon_{zr} = \frac{1}{2}d_r v(r) = -\frac{\Delta p r}{4L\eta}\,, \tag{3.14}$$

and from (3.13)

$$\sigma = \frac{1}{T_0}\frac{\Delta p^2 r^2}{4L^2\eta}\,. \tag{3.15}$$

The total entropy production rate is given as

$$\Sigma = \int_0^L dz \int_0^{2\pi} d\varphi \int_0^R r dr\, \sigma = \frac{\Delta p^2 \pi R^4}{8\eta L T_0}\,.$$

It is interesting to relate the entropy production to the pumping power P_p,

$$P_p = (\delta V/\delta t)\Delta p$$

with the volume δV transported through the pipe in time δt:

$$\delta V/\delta t = \int_0^{2\pi} d\varphi \int_0^R r dr\, v(r)$$

giving

$$P_p = \frac{\pi \Delta p^2 R^4}{8L\eta}$$

and

$$\Sigma = \frac{P_p}{T_0} . \tag{3.16}$$

The whole work needed to maintain the pressure gradient along the pipe is irreversibly transformed into heat, and as a consequence increases the entropy of the heat bath. If we express Σ by the maximal velocity

$$v_m = v(0) = \frac{R^2 \Delta p}{4L\eta}, \tag{3.17}$$

we find

$$\Sigma = \frac{2\pi L \eta v_m^2}{T_0} , \tag{3.18}$$

independent from the pipe's diameter. Taking water with a flow rate of 1 m/s at T_0 = 300 K and L = 10 m, this yields the entropy production rate

$$\Sigma \approx 0.21 \cdot 10^{-3}\,\mathrm{J/Ks} .$$

(b) Non-isothermal case. Friction inside the pipe flow produces heat, and as a consequence the temperature inside the pipe will increase, becoming space-dependent. A steady temperature distribution must be a solution of the stationary heat equation (3.9),

$$0 = \lambda \frac{1}{r}\frac{d}{dr}\left(r \frac{d}{dr} T(r) \right) + \frac{4\eta v_m^2}{R^4}\, r^2 , \tag{3.19}$$

where we used (3.14) and (3.17). Integrating (3.19) twice and assuming the boundary condition $T(R) = T_0$, the temperature of the surrounding heat bath, we get

$$T(r) = T_0 + \Delta T \left(1 - \left(\frac{r}{R} \right)^4 \right) \tag{3.20}$$

with

$$\Delta T = \frac{\eta v_m^2}{4\lambda} \tag{3.21}$$

as the maximal temperature difference of the flow (in the middle) and the pipe wall. Taking again water with v_m = 1 m/s, this gives the rather small value of

$$\Delta T \approx 0.5 \cdot 10^{-3}\,\mathrm{K} .$$

The local entropy density production rate reads now

$$\sigma = \lambda \left(\frac{\nabla T(r)}{T(r)} \right)^2 + \frac{16\lambda r^2}{R^4} \frac{\Delta T}{T(r)}, \tag{3.22}$$

where the last term is equal to (3.15) for variable $T(r)$ if Δp is expressed via (3.17) and (3.21) by ΔT. Computing now the total entropy production rate yields

$$\Sigma = \int_V \sigma \, d^3\vec{r} = 2\pi L \int_0^R \sigma r dr = 8\pi L\lambda \frac{\Delta T}{T_0} = \frac{P_p}{T_0}.$$

This is exactly the same value found for the isothermal case (3.16). Why? The local entropy production rate differs (slightly) if the fluid heats up due to viscous friction; see Figure 3.3 for a highly overdone ΔT. But for the total entropy production it makes no difference if the fluid is assumed isothermal or not. At the end, the whole pumping power goes to the heat bath having the temperature of the pipe wall. Thus the increment of the entropy of the heat bath per time must always be given by P_p/T_0.

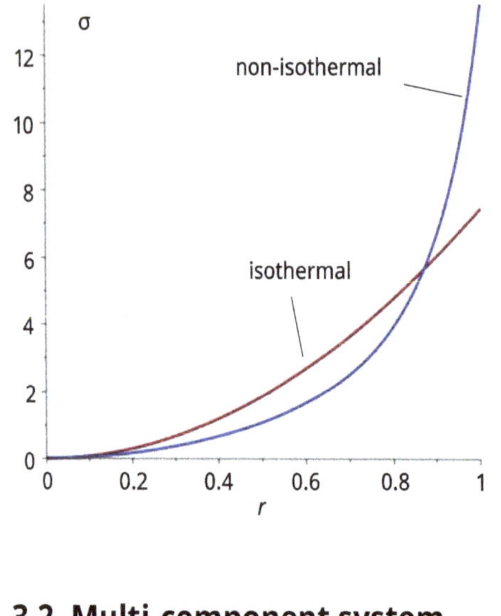

Figure 3.3: Local entropy production $\sigma(r)$ for the isothermal case (red) from Section 3.1.5 (a), equation (3.15), and for the non-isothermal case (blue), equation (3.22) for $T = 300$ K, $R = 1$ m, $\Delta T = 250$ K. For a realistic pipe flow, ΔT would be much smaller, and the difference between the two lines becomes negligible.

3.2 Multi-component system

For a multi-component mixture, the conservation equations for the total mass (3.1) and momentum (3.3) remain unchanged. In addition, one has the equations for the specific densities (2.47) with the sources (2.55) from possible chemical reactions.

3.2.1 Mass fluxes

The remaining task is to close the system by computing $\vec{J}_j$ from the closure (2.77) and to complete $\vec{j}^Q$ along (2.76); we do so now. The chemical potentials can be expressed as functions of p, T and the c_j so that

$$\nabla\left(\frac{\hat{\mu}_j}{T}\right) = \frac{\partial}{\partial T}\left(\frac{\hat{\mu}_j}{T}\right)\nabla T + \frac{\partial}{\partial p}\left(\frac{\hat{\mu}_j}{T}\right)\nabla p + \sum_i^k \frac{\partial}{\partial c_i}\left(\frac{\hat{\mu}_j}{T}\right)\nabla c_i \,. \tag{3.23}$$

The first term on the r. h. s. is converted as

$$\frac{\partial}{\partial T}\left(\frac{\hat{\mu}_j}{T}\right) = -\frac{1}{T^2}\hat{\mu}_j + \frac{1}{T}\frac{\partial \hat{\mu}_j}{\partial T} = -\frac{1}{T^2}(\hat{\mu}_j + T\hat{s}_j) = -\frac{\hat{h}_j}{T^2}\,, \tag{3.24}$$

where we used (1.41)

$$\frac{\partial \hat{\mu}_j}{\partial T} = \frac{\partial^2 \hat{g}}{\partial T \partial c_j} = -\frac{\partial \hat{s}}{\partial c_j} = -\hat{s}_j,$$

and $\hat{h}_j = \hat{g}_j + T\hat{s}_j = \hat{\mu}_j + T\hat{s}_j$ denotes the specific enthalpy.

The second term on the r. h. s. of (3.23) is written as

$$\frac{\partial}{\partial p}\left(\frac{\hat{\mu}_j}{T}\right) = \frac{1}{T}\frac{\partial \hat{\mu}_j}{\partial p} = \frac{1}{T}\frac{\partial^2 \hat{g}}{\partial p \partial c_j} = \frac{1}{T}\frac{\partial \hat{v}}{\partial c_j} = \frac{\hat{v}_j}{T}\,. \tag{3.25}$$

For both conversions, we used the decomposition

$$\hat{\mu}_j = \frac{\partial \hat{g}}{\partial c_j}\,,$$

which follows from (1.47),

$$\hat{g} = \sum_i^k c_i \hat{g}_i = \sum_i^k c_i \hat{\mu}_i$$

and (1.49),

$$\sum_i^k c_i \frac{\partial \hat{\mu}_i}{\partial c_j} = 0\,. \tag{3.26}$$

Since V and S are extensive,

$$\hat{v} = \sum_i^k c_i \hat{v}_i\,, \quad \hat{s} = \sum_i^k c_i \hat{s}_i$$

also applies. With (3.23), (3.24), and (3.25), the mass fluxes (2.77) take the form

$$\vec{j}_i = \sum_j^k \frac{\ell_{ij}}{T} \frac{\vec{f}_j}{\rho_j} - b_i \nabla T - d_i \nabla p - \sum_j^k g_{ij} \nabla c_j \tag{3.27}$$

with the transport coefficients

$$b_i \equiv \frac{\ell_i'}{T^2} - \sum_j^k \frac{\ell_{ij}\hat{h}_j}{T^2}, \quad d_i \equiv \sum_j^k \frac{\ell_{ij}\hat{v}_j}{T}, \quad g_{ij} \equiv \sum_n^k \frac{\ell_{in}}{T} \frac{\partial \hat{\mu}_n}{\partial c_j}. \tag{3.28}$$

In a similar way, the heat flux (2.76) has to be extended:

$$\vec{j}^Q = \sum_j^k a_j \frac{\ell_j}{T} \frac{\vec{f}_j}{\rho_j} - \tilde{\lambda} \nabla T - d \nabla p - \sum_j^k g_j \nabla c_j \tag{3.29}$$

with the coefficients

$$\tilde{\lambda} \equiv \lambda - \sum_j^k \frac{\ell_j \hat{h}_j}{T^2}, \quad d \equiv \sum_j^k \frac{\ell_j \hat{v}_j}{T}, \quad g_j \equiv \sum_i^k \frac{\ell_i}{T} \frac{\partial \hat{\mu}_i}{\partial c_j}. \tag{3.30}$$

3.2.2 Concentration and energy equations

Inserting the fluxes (3.27) into (2.47) yields the rate equations for the specific concentrations

$$\rho \frac{dc_i}{dt} = \sum_j^k g_{ij} \nabla^2 c_j + b_i \nabla^2 T + d_i \nabla^2 p - \sum_j^k \frac{\ell_{ij}}{T} \nabla \cdot \left(\frac{\vec{f}_j}{\rho_j} \right) + q_i. \tag{3.31}$$

Here we assume that the transport coefficients do not depend on space. In fact, they of course do, since they depend on T, p, and c_j. In this way, (3.31) can be considered as linearization with respect to ∇c_j, ∇T, and ∇p. This should be kept in mind when applying the rate equations in nonlinear theories.

Note also that the c_j are not independent on each other, since one has

$$\sum_j^k c_j = 1,$$

and therefore

$$c_j = 1 - \sum_{i \neq j}^k c_i$$

for every $j = 1..k$. Also the relations

$$\sum_j^k \frac{dc_j}{dt} = 0, \quad \sum_j^k \nabla c_j = 0, \quad \sum_j^k \nabla^2 c_j = 0$$

must hold. For the energy equation, we obtain with the heat flux (3.29) the extended form

$$\rho \frac{d\hat{u}}{dt} = \tilde{\lambda}\nabla^2 T + \sum_j^k g_j \nabla^2 c_j + d\nabla^2 p$$
$$- \sum_j^k \frac{\ell_j}{T} \nabla \cdot \left(\frac{\vec{f}_j}{\rho_j}\right) - p\,\nabla \cdot \vec{v} + 2\eta\,\underline{\epsilon} : \underline{\epsilon} + \left(\zeta - \frac{2}{3}\eta\right)(\nabla \cdot \vec{v})^2 . \qquad (3.32)$$

In the following three chapters, we shall discuss various applications of the nonlinear partial differential equations (3.1), (3.3), (3.31), and (3.32).

3.3 Problem: Couette flow

Consider a viscous fluid between two horizontal rigid plates having the distance d; Figure 3.4. If the upper plate moves with a constant velocity U_0 to the right side, the flow between the two plates has the form of a stationary Couette flow,

$$\vec{v} = \frac{z}{d}\, U_0\, \hat{e}_x . \qquad (1)$$

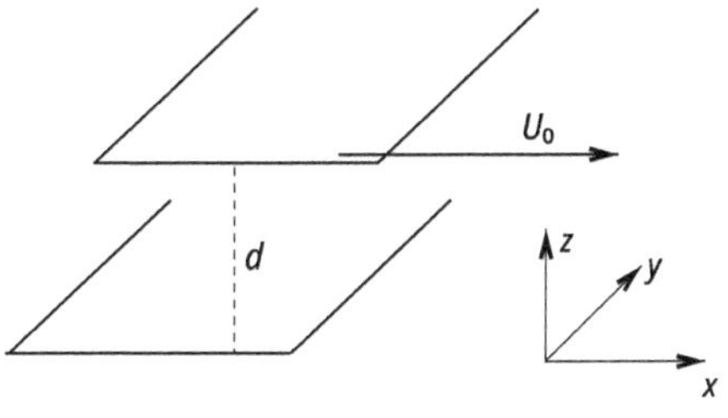

Figure 3.4: A fluid between to infinitely extended horizontal plates, where the upper plate moves with a constant velocity in x-direction.

a) Show that (1) describes an incompressible fluid, and that it is a solution of the Navier–Stokes equations (3.4).
Show also that (1) fulfills the no-slip boundary conditions.
What do you find for the pressure p if $\vec{f} = -\rho g\, \hat{e}_z$?
b) Now let the plates be in contact with a heat bath with constant temperature T_0. Compute the stationary temperature distribution $T(z)$ inside the fluid. Where and how large is the maximal temperature? Compute the temperature difference between the plates and the hottest point for water ($\nu = 10^{-6}\,\text{m}^2/\text{s}$, $\rho = 1000\,\text{kg/m}^3$,

$\lambda = 0.5\,\mathrm{J/Ksm}$) and $U_0 = 1\,\mathrm{m/s}$. How fast must the plate move to obtain a difference of 1 K?

c) Compute the entropy production rate $\sigma(z)$ for the case of an isothermal fluid ($T = T_0$) (neglect heat produced by friction). Compute the total production rate per area,

$$\frac{\Sigma}{A} = \int_0^d dz\, \sigma(z)\,.$$

Show that Σ is equal to the power needed to move the upper plate divided by T_0.

d) Compute the entropy production rate $\sigma(z)$ with $T(z)$ from b). Show that the total rate Σ is the same than that found in c). Why?

4 Diffusion

In this chapter, we consider a medium in rest, $\vec{v} = 0$, of either one component or later of a mixture of two miscible components. In the first case, only heat may diffuse by heat flux; for the latter situation, also one component can migrate into the other through diffusive mass flux.

4.1 Heat equation

We begin with a non-moving one-component system in a given volume, described completely by its temperature distribution $T(\vec{r}, t)$ as a solution of the heat equation (3.9),

$$\boxed{\partial_t T(\vec{r}, t) = \kappa \nabla^2 T(\vec{r}, t)}\,, \tag{4.1}$$

where we assumed a constant thermal diffusivity $\kappa > 0$. Equation (4.1) is a parabolic linear partial differential equation and must be completed by boundary conditions specifying T or ∇T (heat flow) or a combination of both along the boundaries of the integration domain,

$$\alpha\, T(\vec{r}, t) + \beta\, \hat{n} \cdot \nabla T(\vec{r}, t) = f(\vec{r}, t), \quad \vec{r} \in \Omega, \tag{4.2}$$

where Ω denotes the boundary walls and $\hat{n}$ their normal vector. Equation (4.2) is the most general form of boundary conditions, named Robin conditions. If $\beta = 0$, they are called Dirichlet conditions; if $\alpha = 0$, they are called Neumann conditions. For $f = 0$, the boundary conditions are homogeneous, and the trivial solution $T = 0$ for (4.1) exists.

But (4.1) is also an initial value problem, and the initial temperature profile

$$T(\vec{r}, t = t_0) = T_0(\vec{r}) \tag{4.3}$$

must be known in the whole domain to allow for a unique solution.

4.1.1 Dirichlet conditions in one dimension

We first consider (4.1) in one dimension $T = T(x, t)$,

$$\partial_t T = \kappa \frac{\partial^2 T}{\partial x^2}, \tag{4.4}$$

which, for instance, describes the temporal evolution of the temperature distribution along an (infinitely) thin rod having the thermal diffusivity κ and the length L. Dirichlet boundary conditions read

$$T(x = 0, t) = T_a, \quad T(x = L, t) = T_b\,;$$

https://doi.org/10.1515/9783110794137-005

the initial condition shall be given as

$$T(x, t = 0) = T_0(x) .$$

4.1.1.1 Analytical solution

For constant T_a, T_b, the ansatz

$$T(x, t) = T_a + (T_b - T_a)\frac{x}{L} + \vartheta(x, t)$$

transforms the inhomogeneous boundary value problem to a homogeneous one

$$\partial_t \vartheta = \kappa \frac{\partial^2 \vartheta}{\partial x^2}$$

with

$$\vartheta(x = 0, t) = \vartheta(x = L, t) = 0 .$$

Decomposition of ϑ into suitable trigonometric functions yields the analytic solution in form of an infinite series,

$$T(x, t) = T_a + (T_b - T_a)\frac{x}{L} + \sum_{n=1}^{\infty} g_n(t) \sin k_n x \tag{4.5}$$

where

$$g_n(t) = A_n \,\mathrm{e}^{-k_n^2 \kappa t}, \quad k_n = n\pi/L .$$

Finally A_n is determined from the initial condition $T = T_0$,

$$A_n = \frac{2}{L} \int_0^L dx\, T_0(x) \sin k_n x - \frac{2}{n\pi}(T_a - (-1)^n T_b) .$$

For $t \to \infty$, we find $g_n \to 0$, and the temperature asymptotically approaches the stationary solution,

$$T_s(x) = T_a + (T_b - T_a)\frac{x}{L} .$$

Note that even the stationary slab is still not in equilibrium and entropy is permanently produced, as already shown in the example in Section 3.1.4.

4.1.1.2 Temperature distribution in an infinite domain

For the case $L \to \infty$ and the Dirichlet conditions $T(x \to \pm\infty, t) = 0$, equation (4.4) is solved by Fourier transformation

$$T(x,t) = \frac{1}{\sqrt{2\pi}} \int_{-\infty}^{\infty} dk\ \tilde{T}(k)\ \mathrm{e}^{ikx-\kappa k^2 t}\,. \tag{4.6}$$

The Fourier transform $\tilde{T}$ is determined from the initial condition

$$\tilde{T}(k) = \frac{1}{\sqrt{2\pi}} \int_{-\infty}^{\infty} dx\ T_0(x)\ \mathrm{e}^{-ikx}\,.$$

Inserting this into (4.6) yields

$$T(x,t) = \frac{1}{2\sqrt{\pi\kappa t}} \int_{-\infty}^{\infty} dx'\ T_0(x')\ \mathrm{e}^{-\frac{(x-x')^2}{4\kappa t}} = \hat{D}\ T_0\,, \tag{4.7}$$

where the integral over k has been evaluated. The integral operator $\hat{D}$ is also called "propagator," because it propagates the initial condition to any desired time. Taking the special sharply localized initial condition

$$T_0(x) = \theta_0 \delta(x - x_0)\,,$$

(4.7) turns into a Gaussian centered at x_0 with increasing width

$$T(x,t) = \frac{\theta_0}{2\sqrt{\pi\kappa t}}\ \mathrm{e}^{-\frac{(x-x_0)^2}{4\kappa t}}\,. \tag{4.8}$$

4.1.2 A simple numerical method

In more than one spatial dimensions and/or for more complicated boundary conditions, analytic solutions can normally not be found, and one must resort to computer solutions. We present a simple finite difference (FD) method, which is of second order in space and time.

We explain the method with a two-dimensional example. The xy-plane is discretized by a quadratic mesh with mesh size h so that

$$T(x,y,t) = T(ih,jh,t) \equiv T_{ij}(t), \quad i,j = 0\ldots N,$$

and

$$h = L/N$$

with $N+1$ as the number of lines and rows and L as the domain length, here a square; see Figure 4.1. Derivatives are then approximated by differential quotients. Especially for the 2D Laplacian, one finds

$$\nabla^2 T(x,y,t)\big|_{x_i y_j} = \frac{T_{i+1,j} + T_{i-1,j} + T_{i,j+1} + T_{i,j-1} - 4T_{i,j}}{h^2} + O(h^2)\,.$$

In this way, (4.1) is converted to a system of $(N+1)^2$ ordinary differential equations for the $(N+1)^2$ variables $T_{ij}(t)$,

$$d_t T_{ij}(t) = \sum_{m,n}^{N} a_{ijmn} T_{mn}(t)\,. \tag{4.9}$$

Dirichlet conditions are transformed to

$$T_{ij} = f(x_i, y_j), \quad i,j \in \Omega\,.$$

For Neumann conditions, virtual lines outside the domain are constructed; compare Figure 4.1. As example, we consider

$$\frac{\partial T}{\partial x} = f(x,y) \quad \text{at } x = 0\,.$$

Discretizing the first derivative

$$\frac{T_{1,j} - T_{-1,j}}{2h} = f(0, y_j), \quad j = 0 \ldots N$$

can be readily solved for the virtual line

$$T_{-1,j} = T_{1,j} - 2hf(0, y_j), \quad j = 0 \ldots N,$$

which is needed in (4.9) when computing $d_t T_{i,0}$.

Time has to be discretized too, which is again achieved by approximating the time derivative by the corresponding differential quotient. As an example, consider the ordinary differential equation

$$d_t x(t) = f(x(t)),$$

which can be integrated to

$$x(t + \Delta t) = x(t) + \int_t^{t+\Delta t} f(x(t))dt\,. \tag{4.10}$$

We abbreviate

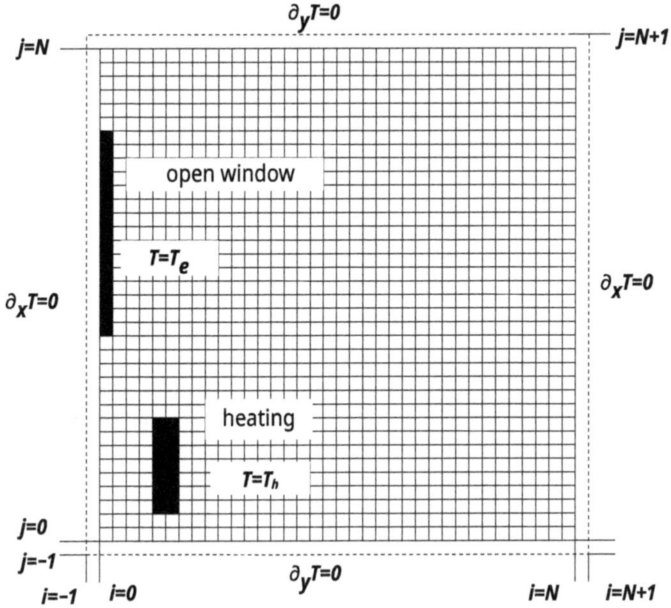

Figure 4.1: Sketch of the system, room with window and heating. $(N+1)\times(N+1)$ FD-grid, dashed: virtual lines. Along window and heater Dirichlet conditions; along walls Neumann conditions apply.

$$x^{(i)} \equiv x(i\Delta t)\,,$$

where Δt is the time step. The integral on the r. h. s. is evaluated numerically replacing f by a polynomial that is fitted to f, taking the n points

$$f(x^{(i+1)}),\ f(x^{(i)}),\ \dots f(x^{(i-n+2)})$$

for an implicit or

$$f(x^{(i)}),\ f(x^{(i-1)}),\ \dots f(x^{(i-n+1)})$$

for an explicit method. We shall concentrate here only on explicit methods. After evaluation of the integral, one finds from (4.10)

$$\begin{aligned}
n=1:&\quad x^{(i+1)} = x^{(i)} + \Delta t\, f_i,\\
n=2:&\quad x^{(i+1)} = x^{(i)} + \frac{\Delta t}{2}(3f_i - f_{i-1}),\\
n=3:&\quad x^{(i+1)} = x^{(i)} + \frac{\Delta t}{12}(23f_i - 16f_{i-1} + 5f_{i-2}),
\end{aligned}$$

where $f_i \equiv f(x^{(i)})$. These are the *Adams–Bashforth* formulas with the discretization error of order Δt^n. The first one is also called *one-step Euler forward method* and yields the lowest accuracy. Adams–Bashforth algorithms are multi-step methods (n steps).

If we take $n = 2$, the system (4.9) turns into the iteration scheme

$$T_{ij}^{(k+1)} = T_{ij}^{(k)} + \frac{\Delta t}{2} \sum_{m,n}^{N} a_{ijmn} (3T_{mn}^{(k)} - T_{mn}^{(k-1)}) . \qquad (4.11)$$

Starting from the initial condition

$$T_{ij}^{(0)} = T(x_i, y_j, t = 0) ,$$

the scheme (4.11) allows for the successive computation of all node values $T_{ij}^{(k)}$ for $k = 2, 3, \dots$. For $k = 1$, the Euler scheme ($n = 1$) has to be used.

Having $T_{ij}^{(k+1)}$ already isolated on the left-hand side, (4.11) is called an *explicit scheme* in time, sometimes also *FTCS scheme*, as acronym for “forward time centered space.” For more details, we refer to [5].

4.1.3 Example: temperature distribution in a room with stove and window

As a first application, we wish to compute the temperature field in a room with a length and height of 3 meters. In the room, a stove with a given temperature T_h is placed below the window. Walls, floor, and ceiling are assumed to be perfect thermal insulators, where homogeneous Neumann boundary conditions $\partial_n T = 0$ apply. Along the open window, the temperature is fixed to the outdoor temperature T_e (Dirichlet); see Figure 4.1. The nodes along Dirichlet boundaries (at the stove and the window) are just fixed at the appropriate values, all other nodes are iterated along equation (4.11). Figure 4.2 shows a time series computed on a 350×350 grid. The step sizes

$$\Delta x = 3/349, \quad \Delta t = 0.1 \Delta x^2 / \kappa$$

have been used, where $\kappa = 2 \cdot 10^{-5} \mathrm{m}^2/\mathrm{s}$ denotes the thermal diffusivity of air. As initial condition, the temperature inside the room was fixed to the outdoor value T_e.

Once the temperature distribution is known, the heat flux according to

$$\vec{j}^Q = -\lambda \nabla T$$

can be computed (λ = thermal conductivity of air) and integrated over the surfaces of window and stove,

$$\langle J_h \rangle = \lambda \left| \int_{F_h} d^2\vec{f} \cdot \nabla T \right| , \quad \langle J_e \rangle = \lambda \left| \int_{F_e} d^2\vec{f} \cdot \nabla T \right| . \qquad (4.12)$$

The temporal evolution of $\langle J_h \rangle$ and $\langle J_e \rangle$ is depicted in Figure 4.3. For the stationary case, both currents should be equal, which is achieved within a tolerance of about 5 percents, that is probably due to numerical errors and the finite resolution.

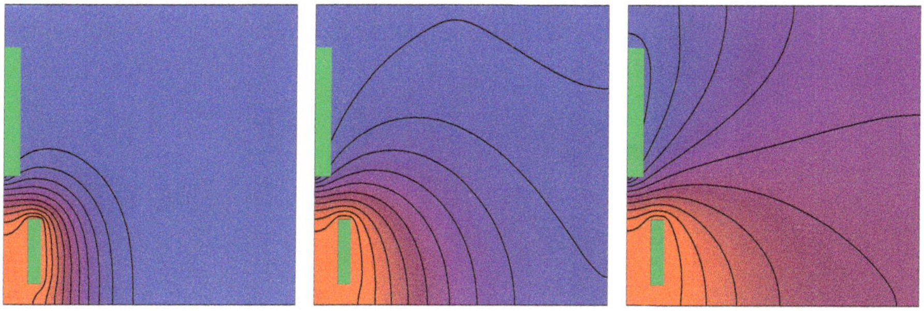

Figure 4.2: Time series, temperature distribution after 0.1, 1, 10 days. Stove and window marked in green, $T_e = 275$ K, $T_h = 345$ K. The extremely long relaxation time is a consequence of the total disregard of any convection.

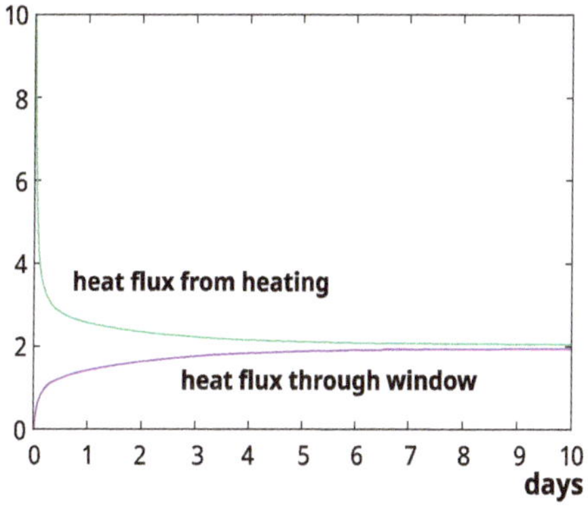

Figure 4.3: Heat flux densities integrated over window and heater in units of J/s as a function of time.

4.1.3.1 Entropy balance

Taking equation (2.38), the total entropy production rate is computed for a given $T(x, y)$ numerically as

$$\Sigma(t) = \lambda \int_V dxdy \, \frac{(\nabla T)^2}{T^2} = \lambda \sum_{ij}^{N} \frac{(T_{i+1,j} - T_{i-1,j})^2 + (T_{i,j+1} - T_{i,j-1})^2}{4T_{ij}^2} \,. \tag{4.13}$$

In the beginning, when the room is cold, Σ is large but decreases monotonically. If everywhere in the room a stationary state $\partial_t T = 0$ is reached, Σ is constant and must be equal to the difference of the entropy fluxes out of the window, and from the heater, that can be computed from (4.12),

$$\frac{\Delta S_{he}}{\Delta t} = \frac{\langle J_e \rangle}{T_e} - \frac{\langle J_h \rangle}{T_h} \,. \tag{4.14}$$

For $\langle J_e \rangle = \langle J_h \rangle$, clearly $\Delta S_{he}/\Delta t > 0$. In the stationary state, (4.14) corresponds to the entropy production rate of the whole system, consisting of the heater (hotter heat bath), the room, and the environment (colder heat bath).

Figure 4.4 shows the temporal evolution of Σ and $\Delta S_{he}/\Delta t$. The small discrepancy for large t is again due to numerical inaccuracy.

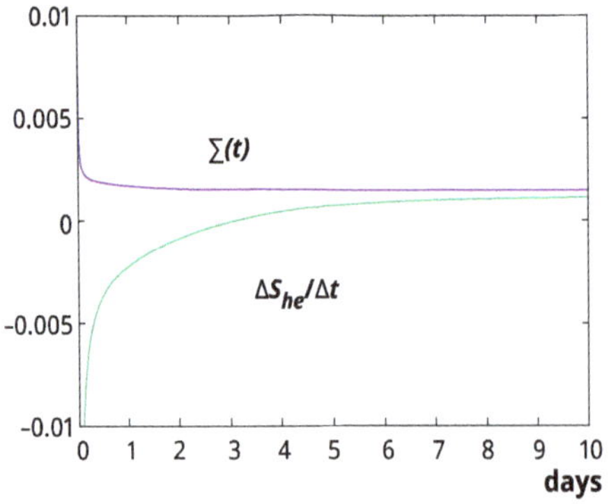

Figure 4.4: Total entropy production rate and entropy change per time of heater and environment in units of J/Ks.

4.2 Total entropy production rate

4.2.1 Linear diffusion equation or linear Onsager coefficients?

In the preceding examples, but already in (3.9), we assumed a linear diffusion equation for the temperature with constant κ, where "constant" means independent from T. This is obtained from Fourier's law (2.68) with constant λ (constant density and specific heat as well). However, taking λ constant leads with (2.67) to a temperature dependent Onsager coefficient $L_v \sim T^2$ and contradicts to the assumption of a linear relation between fluxes and forces. Taking on the other hand a constant L_v renders the temperature equation (3.9) nonlinear,

$$c\rho \frac{\partial T}{\partial t} = -\nabla \cdot L_v \nabla\left(\frac{1}{T}\right) = \frac{L_v}{T^2}\,\nabla^2 T - \frac{2L_v}{T^3}\,(\nabla T)^2 \tag{4.15}$$

(we still do not consider a convecting medium). Of course the assumption $\lambda \sim T^{-2}$ as a consequence of a constant L_v does not correspond to reality.[1] A linear relation, such as (3.9), can be found approximately if we assume T not differing too much from a certain mean temperature T_0,

$$T(\vec{r}, t) = T_0 + \theta(\vec{r}, t) \tag{4.16}$$

with $|\theta| \ll T_0$. Inserting this into (4.15) yields

1 In the last part of this book we shall, based on statistical grounds, derive $\lambda \sim \sqrt{T}$, at least for a perfect gas.

$$\frac{\partial\theta}{\partial t} = \kappa\nabla^2\theta + O(\theta^2) \tag{4.17}$$

with

$$\kappa = \frac{L_v}{c\rho T_0^2},$$

which can now be assumed as a constant. We see that a linear Onsager relation leads to a nonlinear heat equation and vice versa.

4.2.2 "Principle" of minimal entropy production

From the second law, we know that entropy reaches a maximum in thermal equilibrium. Looking at Figure 4.4, one could infer that the entropy production Σ reaches a minimum at the stationary state. This stationary state can be either at equilibrium, then $\Sigma = 0$, or at non-equilibrium with $\Sigma = \text{const}$, as in Figure 4.4. Analogue to the maximum entropy principle, the second law, Prigogine and Glansdorff (1954) formulated the principle of minimal entropy production of a stationary state. This principle, however, holds only for non-moving materials (no convection) and constant Onsager coefficients. It can be shown straightforwardly for the case of pure heat diffusion, as in the example of Section 4.1.3. We assume

$$\vec{j}^Q = L_v\,\nabla\left(\frac{1}{T}\right) \tag{4.18}$$

with constant L_v and consider the total entropy production

$$\Sigma[T] = \int_V d^3\vec{r}\,\vec{j}^Q\cdot\nabla\left(\frac{1}{T}\right) = L_v\int_V d^3\vec{r}\left(\nabla\left(\frac{1}{T}\right)\right)^2 \tag{4.19}$$

as a functional of $T(\vec{r},t)$. The minimum of Σ is found putting the variational derivative to zero, leading to

$$\nabla^2\left(\frac{1}{T}\right) = 0. \tag{4.20}$$

With (4.18), this yields

$$\nabla\cdot j^Q = 0$$

or, due to (2.23c) with $\hat{u} = cT$, $q_U = 0$,

$$c\rho\frac{\partial T}{\partial t} = -\nabla\cdot\vec{j}^Q, \tag{4.21}$$

and finally

$$\frac{\partial T}{\partial t} = 0\,.$$

Thus the solution of (4.20) corresponds to a stationary temperature profile and minimizes entropy production. However, taking instead of (4.18) Fourier's law

$$\vec{j}^Q = -\lambda\,\nabla T$$

with constant λ, (4.19) turns into

$$\Sigma[T] = \lambda \int_V d^3\vec{r}\,\left(\frac{\nabla T}{T}\right)^2 \tag{4.22}$$

with the variational derivative

$$\frac{\delta \Sigma}{\delta T} = \frac{2\lambda}{T^2}\left(\frac{(\nabla T)^2}{T} - \nabla^2 T\right).$$

Clearly, this does not vanish for the stationary state, which is now given from $\nabla^2 T = 0$. Again one can linearize according to (4.16) and may write

$$\frac{\delta \Sigma}{\delta T} = -\frac{2\lambda}{T_0^2}\,\nabla^2\theta + O(\theta^2),$$

which is then zero, at least in linear order of θ. So only for the case of the rather artificial thermal conductivity $\lambda \sim T^{-2}$ and vanishing convection Prigogine's principle can be proved. For other more complicated (nonlinear) cases, it does not hold.

4.2.3 Lyapunov function

For the case of constant L_v, one can even show that $\Sigma(t)$ is monotonically decreasing from every arbitrary initial condition until a stationary state is reached. Forming the time derivative of (4.19), we obtain

$$\frac{d\Sigma}{dt} = -2L_v \int_V d^3\vec{r}\,\nabla\left(\frac{1}{T}\right)\cdot\nabla\left(\frac{\partial_t T}{T^2}\right) = -2\int_V d^3\vec{r}\,\vec{j}^Q\cdot\nabla\left(\frac{\partial_t T}{T^2}\right),$$

where we used (4.18). Now we may integrate in parts

$$\frac{d\Sigma}{dt} = 2\int_V d^3\vec{r}\,(\nabla\cdot\vec{j}^Q)\,\frac{\partial_t T}{T^2} - 2j_n^Q\,\frac{\partial_t T}{T^2}\bigg|_S\,,$$

where the last term denotes surface contributions, which are zero when the temperature is assumed to be constant in time along the boundaries or if the heat flux j_n^Q perpendicular to the boundaries is zero (closed system). With (4.21), we obtain

$$\frac{d\Sigma}{dt} = -2 \int_V d^3\vec{r}\; c\rho\, \frac{(\partial_t T)^2}{T^2} \le 0,$$

because all quantities under the integral are larger or equal to zero.

A function that decreases monotonically under a given dynamics is called *Lyapunov function*. If this function is bounded, it asymptotically reaches a minimum at the stationary state. There can be secondary minima as well; so let us denote the main minima with "ground state." Then the existence of a Lyapunov function proves that this ground state must be stable. Every small perturbation of the ground state leads to a decreasing Σ back to the undisturbed state. Therefore, no bifurcations or instabilities can be expected under the linear assumptions made in (4.18) and in the absence of convection. However, as we will see in Chapters 5 and 6, $\Sigma(t)$ can increase as well and can become periodic, or even chaotic if these conditions are released.

4.2.4 Numerical solutions

We conclude this section presenting numerical simulations for both cases, (i) $L_\nu = \text{const.}$, $\lambda = L_\nu/T^2$, and (ii) $\lambda = \text{const.}$, $L_\nu = \lambda T^2$, discussed in Section 4.2.1. To this end, we integrate the one-dimensional heat equations applying finite differences:

$$\partial_t T = \frac{1}{T^2}\,\partial^2_{xx} T - \frac{2}{T^3}\,(\partial_x T)^2 \quad \text{for (i)}, \tag{4.23a}$$

$$\partial_t T = \partial^2_{xx} T \quad \text{for (ii)} \tag{4.23b}$$

in non-dimensional form. For both cases, we take the same initial condition

$$T(x,0) = 2\xi(x) + 1$$

with $\xi(x)$ equally random distributed in [0,1]. The Dirichlet conditions

$$T(0,t) = 1, \quad T(1,t) = 2$$

are chosen for the walls. The computation is terminated when the mean absolute change of T per time step falls below 10^{-7}. Figure 4.5, left frame, shows the profiles at $t = t_0 = 0.002$ (dashed) and the stationary states at $t = t_1 = 60$. A straight line is obtained for (ii), as expected from the result of Section 4.1.1.1. For (i) the stationary solution looks different due to the temperature dependent thermal conductivity.

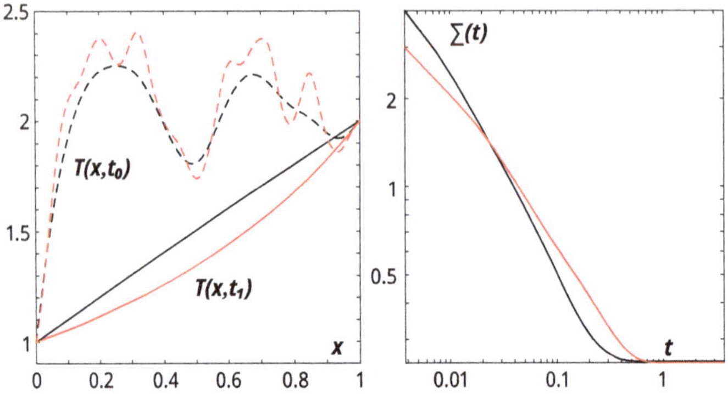

Figure 4.5: Left: temperatures at $t = 0.002$ (dashed) and $t = 60$ computed with (4.23a) (red) and (4.23b) (black). For constant λ, a straight line occurs as expected. Right: Total entropy production rate over time for both cases.

In the right frame of Figure 4.5, we show the entropy production over time, computed with (4.19) for (i) and with (4.22) for (ii). For both cases, a stationary value is obtained after $t \approx 1$. The temporal evolution however looks different.

To sum up, we may say that if the temperature differences are not too large compared to the mean, it is better to take a constant value for λ. If, on the other hand, $\lambda(T)$ (as well as $\rho(T)$ and $c(T)$) play a significant role, one should resort to a nonlinear heat equation of the form

$$\frac{\partial T}{\partial t} = \frac{1}{c(T)\rho(T)}\left[\lambda(T)\,\nabla^2 T + \frac{d\lambda}{dT}\,(\nabla T)^2\right]. \tag{4.24}$$

Of course, the linear Onsager relation between heat flux and force then is only valid up to the first order of ∇T.

4.3 Mass diffusion and thermal diffusion

We wish to compute the rate equations (3.31) for a mixture of two components (gases or arbitrarily miscible fluids). We further assume vanishing forces and no chemical reactions

$$\vec{f}_i = 0, \quad q_i = 0, \quad i = 1, 2\,.$$

We also approximate $p = \text{const.}$, which is normally a good approximation for liquids.

4.3.1 Concentration

Since $c_2 = 1 - c_1$, there exists only one independent concentration, say c_1 and the temperature as state variables. We rename

$$c_1 \equiv c, \quad c_2 = 1 - c . \tag{4.25}$$

We can directly use the balance equation (2.47)

$$\rho \frac{dc}{dt} = -\nabla \cdot \vec{J}_1 \tag{4.26}$$

and $\vec{J}_1$ from (2.77)

$$\vec{J}_i = \ell_i' \, \nabla\left(\frac{1}{T}\right) - \ell_{i1} \nabla\left(\frac{\hat{\mu}_1}{T}\right) - \ell_{i2} \nabla\left(\frac{\hat{\mu}_2}{T}\right)$$

$$= -\frac{1}{T^2}(\ell_i' - \ell_{i1}\hat{h}_1 - \ell_{i2}\hat{h}_2)\nabla T - \left(\frac{\ell_{i1}}{T}\frac{\partial\hat{\mu}_1}{\partial c} + \frac{\ell_{i2}}{T}\frac{\partial\hat{\mu}_2}{\partial c}\right)\nabla c , \tag{4.27}$$

where we used (3.23) with $\nabla p \approx 0$. From $\vec{J}_1 = -\vec{J}_2$, it follows

$$\ell_{11} = -\ell_{21}, \quad \ell_{22} = -\ell_{12}, \quad \ell_1' = -\ell_2'$$

and, due to Onsager's relations, $\ell_{12} = \ell_{21}$. Thus we have the matrix

$$\underline{\ell} = \ell_{11} \begin{pmatrix} 1 & -1 \\ -1 & 1 \end{pmatrix} . \tag{4.28}$$

Inserting (4.27) into (4.26) yields with (4.28)

$$\frac{dc}{dt} = D_s\nabla^2 T + \frac{\ell_{11}}{\rho T}\frac{\partial(\hat{\mu}_1 - \hat{\mu}_2)}{\partial c}\nabla^2 c, \tag{4.29}$$

with the cross diffusion coefficient ($\ell_1 = \ell_1'$)

$$D_s = \frac{\ell_1 - \ell_{11}\Delta\hat{h}}{T^2\rho} \tag{4.30}$$

and

$$\Delta\hat{h} = \hat{h}_1 - \hat{h}_2 .$$

To evaluate the last term on the r.h.s. of (4.29), we use the Duhem–Margule relation (3.26) with $j = 1$:

$$\frac{\partial(\hat{\mu}_1 - \hat{\mu}_2)}{\partial c} = \frac{\partial\hat{\mu}_1}{\partial c_1} + \frac{c_1}{c_2}\frac{\partial\hat{\mu}_1}{\partial c_1} = \frac{1}{1-c}\frac{\partial\hat{\mu}_1}{\partial c} \tag{4.31}$$

and obtain from (4.29)

$$\frac{dc}{dt} = D\nabla^2 c + D_s\nabla^2 T \tag{4.32}$$

with the diffusion constant

$$D = \frac{\ell_{11}}{\rho T(1-c)} \frac{\partial \hat{\mu}_1}{\partial c} . \tag{4.33}$$

4.3.2 Temperature

To derive an equation for the temperature field, we use equation (3.32) with $\nabla^2 p \approx 0$, $\vec{f}_j = 0$ and replace $\nabla \cdot \vec{v} = \rho d\hat{v}/dt$ (continuity equation),

$$\rho \frac{d\hat{u}}{dt} = \tilde{\lambda}\nabla^2 T + g_1 \nabla^2 c - p\rho \frac{d\hat{v}}{dt} \tag{4.34}$$

where we have neglected the nonlinear viscous friction terms, and the coefficient g_1 is defined in (3.30). If pressure is one of the independent variables, it is convenient to use the specific enthalpy $\hat{h}$,

$$\hat{u} = \hat{h} - p\hat{v}, \quad du = dh - pd\hat{v} - \hat{v}dp$$

instead of internal energy and replace

$$\frac{d\hat{u}}{dt} = \frac{d\hat{h}}{dt} - p\frac{d\hat{v}}{dt} - \hat{v}\frac{dp}{dt}$$

in (4.34), giving

$$\rho \frac{d\hat{h}}{dt} = \tilde{\lambda}\nabla^2 T + g_1 \nabla^2 c, \tag{4.35}$$

where we again neglect pressure changes. The enthalpy is a function of T, p, c and substituting ($dp \approx 0$)

$$d\hat{h} = \frac{\partial \hat{h}}{\partial T} dT + \sum_i \frac{\partial \hat{h}}{\partial c_i} dc_i$$

into (4.35) yields with $\partial \hat{h}/\partial T = c_p$

$$\rho c_p \frac{dT}{dt} = \tilde{\lambda}\nabla^2 T + g_1 \nabla^2 c - \rho \sum_i \hat{h}_i \frac{dc_i}{dt} , \tag{4.36}$$

where we used $\partial \hat{h}/\partial c_i = \hat{h}_i$. The last term in (4.36) can be written as

$$\rho \sum_i \hat{h}_i \frac{dc_i}{dt} = \rho \Delta \hat{h} \frac{dc}{dt} = \rho \Delta \hat{h}(D\nabla^2 c + D_s \nabla^2 T) ,$$

where we used (4.32) for the time derivative. Finally, equation (4.36) takes the form

$$\frac{dT}{dt} = \tilde{\kappa}\nabla^2 T + D_d \nabla^2 c, \tag{4.37}$$

where

$$\tilde{\kappa} = \frac{\tilde{\lambda}}{c_p \rho} - \frac{\Delta\hat{h}}{c_p} D_s, \quad D_d = \frac{g_1}{c_p \rho} - \frac{\Delta\hat{h}}{c_p} D\,. \tag{4.38}$$

From (3.30), we evaluate with (4.31)

$$g_1 = \frac{\ell_1}{T}\frac{\partial(\hat{\mu}_1 - \hat{\mu}_2)}{\partial c} = \frac{\ell_1}{Tc_2}\frac{\partial\hat{\mu}_1}{\partial c}$$

and

$$\tilde{\lambda} = \lambda - \frac{\ell_1 \Delta\hat{h}}{T^2}\,.$$

4.3.3 The system

We repeat the basic system for $c(\vec{r}, t)$ and $T(\vec{r}, t)$:

$$\boxed{\begin{aligned} \frac{dc}{dt} &= D\,\nabla^2 c + D_s\,\nabla^2 T \\ \frac{dT}{dt} &= \tilde{\kappa}\,\nabla^2 T + D_d\,\nabla^2 c \end{aligned}}\,. \tag{4.39}$$

This is a set of partial differential equations, which must be completed by boundary conditions and initial conditions originating from the physics of the system under consideration. The coefficients in (4.39) are given in (4.30), (4.33) and (4.38) as

$$D = \frac{\ell_{11}}{\rho T(1-c)}\frac{\partial\hat{\mu}_1}{\partial c}, \tag{4.40a}$$

$$D_s = \frac{\ell_1 - \ell_{11}\Delta\hat{h}}{T^2 \rho}, \tag{4.40b}$$

$$\tilde{\kappa} = \kappa_p - \frac{\Delta\hat{h}(2\ell_1 - \ell_{11}\Delta\hat{h})}{T^2 c_p \rho}, \tag{4.40c}$$

$$D_d = \frac{\ell_1 - \ell_{11}\Delta\hat{h}}{T(1-c)c_p\rho}\frac{\partial\hat{\mu}_1}{\partial c} \tag{4.40d}$$

with

$$\kappa_p = \frac{\lambda}{\rho c_p}$$

as the thermal diffusivity of the one-component system. It is clear that all the coefficients (4.40) depend on temperature, pressure, and concentration. However, for the derivation they were assumed being constant, and T, p, c have to be replaced by their mean values T_0, p_0, c_0, from which the deviations should be small. Then (4.39) can be considered as a linearization around T_0, p_0, c_0.

We see that there are only three phenomenological coefficients left, namely

$$\kappa > 0, \quad \ell_1, \quad \ell_{11} > 0,$$

which have to be determined experimentally or from more fundamental theories, including special material properties.

Can D be negative? Taking the Gibbs free energy, we may write according to (1.46)

$$\frac{\partial \hat{\mu}_i}{\partial c_j} = \frac{\partial^2 \hat{g}}{\partial c_j \partial c_j} .$$

On the other hand, a thermodynamic system is in a stable thermal equilibrium if $\hat{g}$ has a minimum. The second variation of $\hat{g}$ reads

$$d^2\hat{g} = \sum_{ij} \frac{\partial^2 \hat{g}}{\partial c_j \partial c_j} dc_i dc_j = \sum_{ij} \frac{\partial \hat{\mu}_i}{\partial c_j} dc_i dc_j .$$

At minimum, $d^2\hat{g} > 0$, and we conclude that the matrix $\{\frac{\partial \hat{\mu}_i}{\partial c_j}\}$ must be positive definite. Therefore,

$$\mathrm{Tr}\left\{\frac{\partial \hat{\mu}_i}{\partial c_j}\right\} = \sum_i \frac{\partial \hat{\mu}_i}{\partial c_i} > 0 .$$

For our two-component system, we may use the Duhem–Margule relations (1.48), (1.49), and write

$$\mathrm{Tr}\left\{\frac{\partial \hat{\mu}_i}{\partial c_j}\right\} = \frac{\partial \hat{\mu}_1}{\partial c_1} + \frac{\partial \hat{\mu}_2}{\partial c_2} = \frac{\partial \hat{\mu}_1}{\partial c_1}\left(1 + \frac{c_1^2}{c_2^2}\right) = \frac{\partial \hat{\mu}_2}{\partial c_2}\left(1 + \frac{c_2^2}{c_1^2}\right) > 0,$$

and from there

$$\frac{\partial \hat{\mu}_i}{\partial c_i} > 0, \quad i = 1, 2 .$$

Thus,

$$D > 0 \tag{4.41}$$

for a stable system.

It is remarkable that the two cross diffusion coefficients (4.40b) and (4.40d) are linked without any phenomenological coefficient according to

$$D_d = \frac{\rho T^2 D}{\ell_{11}} D_s = \frac{D_s T}{(1-c)c_p} \frac{\partial \hat{\mu}_1}{\partial c} . \tag{4.42}$$

Due to (4.41), D_d and D_s must always have the same sign. But there is still one more condition, which comes from the requirement of positive semi-definiteness of the Onsager matrix (second law) introduced in (2.62), which reads in this case

$$\underline{O} = \begin{pmatrix} L_v & \ell_1 \\ \ell_1 & \ell_{11} \end{pmatrix} .$$

Clearly, $\mathrm{Tr}\,\underline{O} > 0$, and from $\mathrm{Det}\,\underline{O} > 0$, it follows

$$\ell_1^2 \le \ell_{11} L_v = \ell_{11} \kappa_p \rho c_p T^2 \tag{4.43}$$

with $L_v > 0$ introduced in (2.66). Inserting (4.43) into (4.40c) finally yields

$$\tilde{\kappa} \ge \frac{(\ell_{11}\Delta\hat{h} - \ell_1)^2}{\ell_{11}\rho T^2} > 0 .$$

4.3.4 Solution: plane waves

We discuss a special solution of (4.39) in one spatial dimension on $0 \le x \le L$. For periodic boundary conditions,[2] we may write

$$c(x,t) = c_0 + \sum_k c_k \mathrm{e}^{ikx+ak^2 t}, \quad T(x,t) = T_0 + \sum_k T_k \mathrm{e}^{ikx+ak^2 t} \tag{4.44}$$

with $k_n = 2\pi n/L$, $n = 1 \ldots \infty$, and T_0, c_0 constant. Inserting into (4.39) yields the solvability condition

$$a^2 + a(D + \tilde{\kappa}) + D\tilde{\kappa} - D_s D_d = 0$$

with roots

$$a = \frac{1}{2}\left(-D - \tilde{\kappa} \pm \sqrt{(D-\tilde{\kappa})^2 + 4D_s D_d}\right) . \tag{4.45}$$

Because of (4.42), D_s and D_d must have the same sign, and therefore a is always real valued. This excludes oscillating waves in (4.44). Moreover, $a \le 0$ as long as

2 Periodic b. c. means physically no boundaries at all, and the system can be considered from the thermodynamic point of view as being closed.

$$D_s D_d \leq D\tilde{\kappa}\,.$$

Inserting the coefficients (4.40) into the last inequality one recovers exactly the condition (4.43). In other words, if the second law is fulfilled, $\alpha < 0$ and

$$\lim_{t\to\infty} c = c_0, \quad \lim_{t\to\infty} T = T_0\,;$$

the perturbations (4.44) decay exponentially in time to global equilibrium, independently from parameters and initial conditions.

4.3.5 Isothermal diffusion

For the isothermal case $T = \text{const.}$ and $\nabla T = 0$, (4.39) reduces to the diffusion equation

$$\partial_t c = D\,\vec{\nabla}^2 c, \tag{4.46}$$

which is also called Fick's (second) law.[3] A plane wave ansatz (4.44) yields

$$\alpha = -D < 0,$$

and all waves decay exponentially. An isothermal situation with $\nabla^2 c \neq 0$ is of course only possible if the cross diffusion coefficient $D_d \approx 0$.

4.3.6 Thermodiffusion and Soret effect

Now we assume a non-isothermal binary mixture in rest in a vertical constant gravity field,

$$\vec{f}_i = -\rho_i g\hat{e}_z, \quad \vec{f} = \vec{f}_1 + \vec{f}_2 = -\rho g\hat{e}_z$$

with $g = 9.81\,\text{m/s}^2$. Due to gravity, the pressure depends on z and from (3.3)

$$d_z p(z) = -\rho g\,. \tag{4.47}$$

From here on, we assume that all variables in the non-moving state depend only on z. Including the pressure variation, the set (4.39) extends to

$$\partial_t c = D\,\partial^2_{zz} c + D_s\,\partial^2_{zz} T + \frac{d_1}{\rho}\partial^2_{zz} p\,, \tag{4.48a}$$

$$\partial_t T = \tilde{\kappa}\,\partial^2_{zz} T + D_d\,\partial^2_{zz} c + \frac{d}{\rho c_p}\partial^2_{zz} p\,. \tag{4.48b}$$

3 Adolf Eugen Fick, German physician, 1829–1901.

With (3.28) and (3.30),

$$d_1 = \frac{\ell_{11}(\hat{v}_1 - \hat{v}_2)}{T}, \quad d = \frac{(\ell_1 - \ell_{11}(\hat{h}_1 - \hat{h}_2))(\hat{v}_1 - \hat{v}_2)}{T},$$

and due to

$$\hat{v} = c_1\hat{v}_1 + c_2\hat{v}_2 = c\hat{v}_1 + (1-c)\hat{v}_2,$$

we find

$$\hat{v}_1 - \hat{v}_2 = \frac{\partial \hat{v}}{\partial c} = \frac{\partial}{\partial c}\frac{1}{\rho} = -\frac{1}{\rho^2}\frac{\partial \rho}{\partial c}$$

and

$$d_1 = -\frac{\ell_{11}}{T}\frac{1}{\rho^2}\frac{\partial \rho}{\partial c} \quad d = -\frac{\ell_1 - \ell_{11}(\hat{h}_1 - \hat{h}_2)}{T}\frac{1}{\rho^2}\frac{\partial \rho}{\partial c}.$$

To continue, we need a state equation for $\rho(T,c)$. We assume a linear dependence

$$\rho(T,c) = \rho_0(1 + \alpha_T(T - T_0) + \alpha_c(c - c_0)) \tag{4.49}$$

with the expansion coefficients

$$\alpha_T = \frac{1}{\rho_0}\left(\frac{\partial \rho}{\partial T}\right)_{c_0,T_0}, \quad \alpha_c = \frac{1}{\rho_0}\left(\frac{\partial \rho}{\partial c}\right)_{c_0,T_0}$$

for heat and concentration, respectively. Using (4.47), we see that the pressure diffusion term in (4.48a)

$$\frac{d_1}{\rho}\partial^2_{zz}p = -\frac{\ell_{11}}{T\rho^3}\frac{\partial \rho}{\partial c}\partial^2_{zz}p = \frac{\ell_{11}g}{T\rho^2}\frac{\partial \rho}{\partial c}\frac{\partial \rho}{\partial z} = \frac{\ell_{11}g}{T}\alpha_c\left(\alpha_c\frac{\partial c}{\partial z} + \alpha_T\frac{\partial T}{\partial z}\right)$$

is of second order in α_T, α_c. The same is valid for the pressure term in (4.48b), and since we are satisfied here with the linear order, we can neglect the pressure diffusion in (4.48a), (4.48b). From here on, we also neglect the at least in liquids normally small coupling from concentration gradient to temperature, and put $D_d = 0$. Then a stationary solution of (4.48b) reads

$$T(z) = T_0 + \beta_T z,$$

where

$$\beta_T = \frac{T_1 - T_0}{L}$$

denotes the temperature gradient in vertical direction, L the depth of the layer, and T_0, T_1 its lower and upper temperatures, respectively. The stationary concentration distribution is then given from (4.48a) by

$$\partial^2_{zz} c = 0$$

and also a linear profile

$$c(z) = c_0 + \beta_c z$$

with

$$\beta_c = \frac{c_1 - c_0}{L} .$$

The two gradients β_T, β_c are connected via the boundary conditions. If we assume vanishing mass flux at the boundaries $z = 0, L$, we have from (4.48a)

$$J_z = -\rho D_s \frac{\partial T}{\partial z} - d_1 \frac{\partial p}{\partial z} - D\rho \frac{\partial c}{\partial z} = 0, \quad z = 0, L$$

or

$$-\rho D_s \beta_T - \frac{g \ell_{11} a_c}{T} - D\rho\beta_c = 0 ,$$

where again only linear terms are kept. Solving the last equation for β_c yields

$$\beta_c = -\frac{D_s}{D} \beta_T - g a_c (1 - c_0) \left(\frac{\partial \hat{\mu}_1}{\partial c} \right)^{-1} .$$

The ratio

$$s_T = \frac{D_s}{D}$$

is called *Soret coefficient*. We see that both gravity and temperature gradient may create a concentration gradient. Gravity leads to a steady configuration, where the more heavy component (if $a_c > 0$ this is the one with $c_1 = c$) tends to accumulate near the ground and the concentration gradient is negative. The mechanism where a temperature gradient accounts for a concentration gradient (first term) is called *Soret effect*. Depending on the sign of D_s, the two gradients are either parallel or reverse (Figure 4.6).

We shall come back to this point in Chapter 6, where we consider a binary mixture in motion.

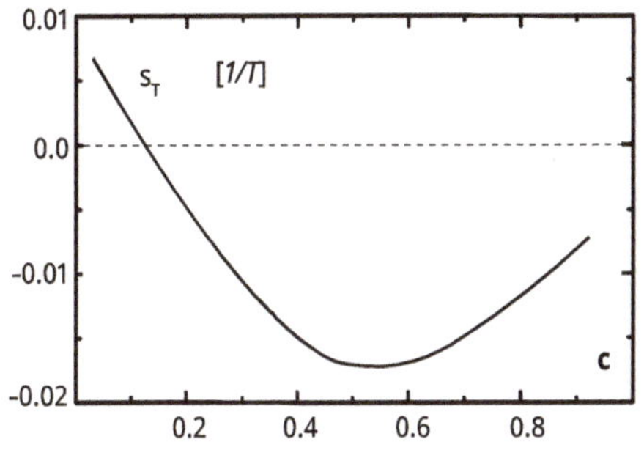

Figure 4.6: Sketch of the Soret coefficient of an acetone/water mixture over the relative acetone concentration. The Soret coefficient depends strongly on the mean concentration and may change sign. Data from H. Ning and S. Wiegand, J. Chem. Phys. 125, 221102 (2006).

4.4 Problem: subterrestrial temperature variations

The daily (or yearly) subterrestrial temperature variations can be approximated by solving the heat equation in one spatial dimension

$$\partial_t T(z,t) = \kappa \partial_{zz} T(z,t)\,, \tag{1}$$

where $z = 0$ denotes the surface of the earth. The temperature must fulfill the boundary conditions (Figure 4.7),

$$T(0,t) = T_0 \cos(\omega t) + T_M \quad \text{at } z = 0, \tag{2}$$

$$T(z,t) = T_M \quad \text{for } z \to -\infty\,. \tag{3}$$

Here, T_M is the constant mean temperature, and $\tau = 2\pi/\omega$ is the length of the periodic variations, i. e., one day or one year.

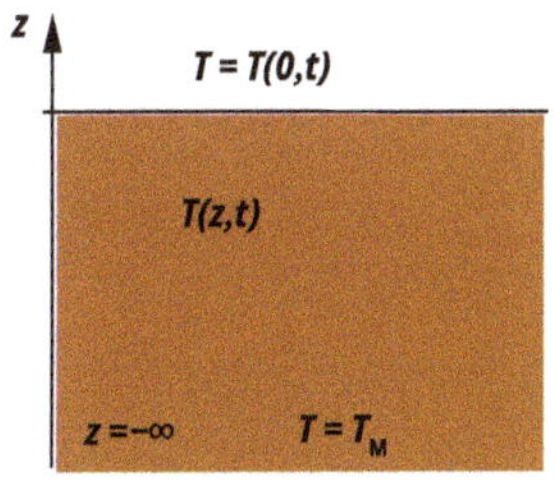

Figure 4.7: Given the temperature at the surface, the subterrestrial temperature variation can be determined solving the 1D heat equation (1).

a) To solve equation (1), a separation ansatz of the form

$$T(z,t) = f(z)\, g(t)$$

is successful. Determine $f(z)$ and $g(t)$. Show that the complete solution of (1) can be written in the form

$$T(z,t) = A\, \mathrm{e}^{\alpha z + \kappa \alpha^2 t} + c.c. + K_0\,, \tag{4}$$

where A, α are complex numbers, $c.c.$ denotes the complex conjugate, and K_0 is some real valued constant.

b) Determine α from the boundary conditions (2), (3), and show that the solution can be expressed as

$$T(z,t) = T_0\,e^{\gamma z}\cos(\omega t + \gamma z) + T_M \tag{5}$$

with

$$\gamma = \sqrt{\frac{\omega}{2\kappa}}$$

as an inverse penetration depth.

c) Discuss the solution (5). Plot T over z for several times. Determine the phase speed. What can you say about daily and yearly penetration depths? Assume $\kappa = 0.6 \cdot 10^{-6}\,\mathrm{m}^2/\mathrm{s}$.

5 Diffusion and reaction

Now we shall allow for chemical reactions and include sources and sinks according to (2.55) into the mass balance equations (2.47)

$$\rho \frac{dc_j}{dt} + \nabla \cdot \vec{J}_j = \sum_r^R \omega_r \, \nu_{rj} M_j \, . \tag{5.1}$$

During this chapter, with the exception of Section 5.1.4, we assume T, p being constant (isothermal-isobar processes).

5.1 Knallgas reaction

We begin with a simple example. The German, also Scandinavian, term "knallgas" describes a mixture of hydrogen and oxygen undergoing an exothermic reaction to water as already discussed in Section 2.3.3, accompanied by a loud crack (knall):

$$2\,H_2 + O_2 \underset{K'}{\overset{K}{\rightleftarrows}} 2\,H_2O \, . \tag{5.2}$$

We assign the mass fractions

$$H_2 \rightarrow c_1, \quad O_2 \rightarrow c_2, \quad H_2O \rightarrow c_3 \, ;$$

the stoichiometric coefficients read

$$\nu_1 = -2, \quad \nu_2 = -1, \quad \nu_3 = 2 \, .$$

There is only one reaction and the reaction velocity is proportional to the possibility that the molecules needed are around and can meet in the same volume element ΔV,

$$\omega = K \tilde{c}_1^2 \tilde{c}_2 - K' \tilde{c}_3^2$$

with the reaction rate constants $K, \; K' \; > \; 0$ and the molar densities $\tilde{c}_i$. Note that the molar densities are related to the mass fractions c_i via

$$\tilde{c}_i = \frac{\rho}{M_i} c_i$$

with the molar mass M_i. Then (5.1) reads

$$\rho \frac{dc_j}{dt} + \nabla \cdot \vec{J}_j = M_j \nu_j \omega, \quad j = 1 \ldots 3, \tag{5.3}$$

https://doi.org/10.1515/9783110794137-006

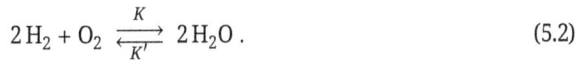

and due to mass conservation

$$c_3 = 1 - c_2 - c_1, \quad 2M_3 = 2M_1 + M_2 \,, \tag{5.4}$$

it is sufficient to consider only two equations, say the first two,

$$\begin{aligned} \rho\partial_t c_1 + \nabla \cdot \vec{j}_1 &= -2M_1\omega, \\ \rho\partial_t c_2 + \nabla \cdot \vec{j}_2 &= -M_2\omega \,, \end{aligned} \tag{5.5}$$

or, expressed in molar densities,

$$\begin{aligned} \partial_t \tilde{c}_1 + \nabla \cdot \vec{\tilde{j}}_1 &= -2(K\tilde{c}_1^2\tilde{c}_2 - K'\tilde{c}_3^2), \\ \partial_t \tilde{c}_2 + \nabla \cdot \vec{\tilde{j}}_2 &= -(K\tilde{c}_1^2\tilde{c}_2 - K'\tilde{c}_3^2) \,, \end{aligned} \tag{5.6}$$

and $\vec{\tilde{j}}_i = \vec{j}_i/M_i$. This system is closed by expressing $\vec{j}_j$ by c_i according to (2.77).

5.1.1 Rate equations

Before we do this, we consider the case where spatial inhomogeneities can be neglected, e. g., in a stirred medium, and find from (5.6) the *rate equations* for $\tilde{c}_i = \tilde{c}_i(t)$,

$$\frac{d\tilde{c}_1}{dt} = -2(K\tilde{c}_1^2\tilde{c}_2 - K'\tilde{c}_3^2), \tag{5.7a}$$

$$\frac{d\tilde{c}_2}{dt} = -(K\tilde{c}_1^2\tilde{c}_2 - K'\tilde{c}_3^2) \,. \tag{5.7b}$$

Eliminating the reaction terms in (5.7) yields

$$\frac{d}{dt}\left(\frac{\tilde{c}_1}{2} - \tilde{c}_2\right) = 0 \,,$$

and from there

$$\tilde{c}_2 = \frac{1}{2}\,\tilde{c}_1 + K_1 \,, \tag{5.8}$$

where the integration constant K_1 is fixed by the initial conditions. For simplicity, we put $K_1 = 0$. Eliminating $\tilde{c}_2$ with (5.8) and using $\tilde{c}_3 = \rho/M_3 - \tilde{c}_1$ (mass conservation (5.4)), (5.7a) reads

$$\frac{d\tilde{c}_1}{dt} = 2K'\alpha^2 - 4K'\alpha\tilde{c}_1 + 2K'\tilde{c}_1^2 - K\tilde{c}_1^3 \tag{5.9}$$

with the abbreviation $\alpha = \rho/M_3$. A stationary equilibrium solution $d_t\tilde{c}_1^s = 0$ can be found approximately for small K' as

$$\tilde{c}_1^s = 2\tilde{c}_2^s = (2\alpha^2)^{1/3}\left(\frac{K'}{K}\right)^{1/3} + O((K'/K)^{2/3}) \,.$$

If we neglect the backwards reaction completely ($K' = 0$),[1] (5.9) simplifies to

$$\partial_t \tilde{c}_1 = -K\tilde{c}_1^3, \tag{5.10}$$

which can be solved analytically with

$$\tilde{c}_1(t) = 2\tilde{c}_2(t) = \frac{\tilde{c}_1(0)}{\sqrt{1 + 2K\tilde{c}_1^2(0)\, t}}$$

with the initial value $\tilde{c}_1(0)$. That means the molar densities of hydrogen and oxygen start at a certain initial value and decrease monotonically until the fixed point $\tilde{c}_1 = \tilde{c}_2 = 0$ is reached and all molecules have been reacted to water.

It is instructive to compute the local entropy production rate for this reaction according to (2.58):

$$\sigma(t) = -\frac{\omega(t)}{T}\sum_{j=1}^{3} \nu_i \hat{\mu}_j M_j = \frac{\omega(t)}{T}\left[M_1(2\hat{\mu}_1 - 2\hat{\mu}_3) + M_2(\hat{\mu}_2 - \hat{\mu}_3)\right]. \tag{5.11}$$

Since $\omega(t)$ goes monotonically to zero, this holds also for σ. However, to have $\sigma > 0$ (2nd law) the condition

$$M_1(2\hat{\mu}_1 - 2\hat{\mu}_3) + M_2(\hat{\mu}_2 - \hat{\mu}_3) > 0 \tag{5.12}$$

must hold, because the reaction is exothermic and takes place in a closed system. Condition (5.12) is clearly fulfilled as long as the total chemical potential of the reactants, $2M_1\hat{\mu}_1 + M_2\hat{\mu}_2$, is larger than that of the product, $2M_3\hat{\mu}_3$, which is just another definition of exothermic. Thus, an exothermic reaction in a closed system produces entropy as long as it is running.

5.1.2 Reaction-diffusion equations

We again eliminate c_3 by (5.4) and write the fluxes as

$$\vec{j}_i = -g_{i,1}\nabla c_1 - g_{i,2}\nabla c_2 + g_{i,3}\nabla(c_1 + c_2) = -\rho D_{i,1}\nabla c_1 - \rho D_{i,2}\nabla c_2, \quad i = 1\ldots 3 \tag{5.13}$$

and with (3.27), (3.28)

$$\rho D_{ij} = g_{ij} - g_{i,3} = \frac{1}{T}\sum_{n}^{3} \ell_{in}\left(\frac{\partial \hat{\mu}_n}{\partial c_j} - \frac{\partial \hat{\mu}_n}{\partial c_3}\right).$$

1 The knallgas reaction is highly exothermic, so putting $K' = 0$ is a quite reasonable approximation.

Due to mass conservation,

$$\sum_i^3 \vec{j}_i = 0\,,$$

and therefore

$$\sum_i^3 D_{ij} = 0, \quad j = 1,2$$

must hold. We have always (2nd law) $D_{ii} > 0$, $i = 1,2$ and very often

$$|D_{ij}| \ll D_{ii}, \quad i \neq j,$$

and we can neglect the cross diffusion terms, giving

$$\vec{j}_i = -\rho D_i \nabla c_i, \quad i = 1,2$$

with the abbreviation $D_i = D_{ii}$. Finally, the reaction-diffusion equations, now for molar densities, for the knallgas reaction ($K' = 0$) read

$$\begin{aligned} \partial_t \tilde{c}_1 &= D_1 \nabla^2 \tilde{c}_1 - 2K \tilde{c}_1^2 \tilde{c}_2, \\ \partial_t \tilde{c}_2 &= D_2 \nabla^2 \tilde{c}_2 - K \tilde{c}_1^2 \tilde{c}_2 \end{aligned} \tag{5.14}$$

with the two diffusion coefficients $D_i > 0$, which have to be determined experimentally or by more fundamental material theories.

5.1.3 Numerical solution

Next we consider the reaction-diffusion equations (5.3) in a one-dimensional reaction chamber of length L and assume $D_1 = D_2 = D_3 = D$. Introducing dimensionless variables by scaling length with L, time with L^2/D, we obtain

$$\begin{aligned} \partial_t \tilde{c}_1 &= \partial^2_{xx} \tilde{c}_1 - 2\tilde{K} \tilde{c}_1^2 \tilde{c}_2, \\ \partial_t \tilde{c}_2 &= \partial^2_{xx} \tilde{c}_2 - \tilde{K} \tilde{c}_1^2 \tilde{c}_2, \\ \partial_t \tilde{c}_3 &= \partial^2_{xx} \tilde{c}_3 + 2\tilde{K} \tilde{c}_1^2 \tilde{c}_2, \end{aligned} \tag{5.15}$$

where x, t are dimensionless with $0 \le x \le 1$ and $\tilde{K} = KL^2/D$. The chamber is supposed to be closed, leading to no-flux boundary conditions

$$\left.\frac{\partial \tilde{c}_i}{\partial x}\right|_{x=0,1} = 0, \quad i = 1,2,3\,.$$

The code is the same as described already in Section 4.1.2. We chose the initial conditions with

$$\tilde{c}_1(x_i, 0) = (0.2 + \xi_i)\ 1/\mathrm{m}^3, \quad \tilde{c}_2(x_i, 0) = (0.1 + \xi_i)\ 1/\mathrm{m}^3,$$

with equally distributed random numbers $-0.05 \le \xi_i < 0.05$. Then the initial mean values are $\langle c_1 \rangle = 2\langle \tilde{c}_2 \rangle$, which ensures no excess of one of the two reactants at any time. Mass conservation yields the condition

$$\tilde{c}_3(x, t) = \frac{\rho}{M_3} - \frac{M_2}{M_3}\,\tilde{c}_2(x, t) - \frac{M_1}{M_3}\,\tilde{c}_1(x, t),$$

which we can use to initialize $\tilde{c}_3$. For the molar masses, we have

$$M_1 = 2\,\mathrm{g}, \quad M_2 = 32\,\mathrm{g}, \quad M_3 = 18\,\mathrm{g}\,;$$

for the density, we take $\rho = 5\,\mathrm{g/m}^3$, and $\tilde{K} = 1$. Here, ρ can be considered as a control parameter and is related to the amount of H_2O present in the beginning.

A time series is shown in Figure 5.1. Diffusion flattens the initial distributions rather fast, and finally a stationary state, which is in equilibrium, is formed when the reactants are consumed completely with

$$\tilde{c}_1^S = \tilde{c}_2^S = 0, \quad \tilde{c}_3^S = \rho/M_3\,.$$

Note that there is still no macroscopic motion (convection) assumed and all fluxes are purely diffusive. Also thermal effects are totally neglected.

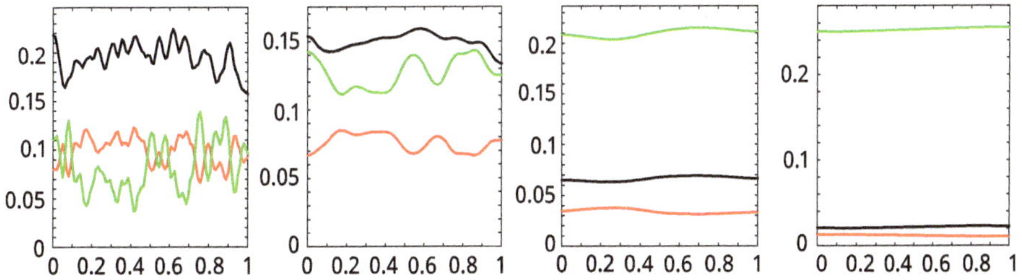

Figure 5.1: Time evolution of the molar densities $\tilde{c}_i(x, t)$ (in units $1/\mathrm{m}^3$) over x at times $t = 1, 10, 100, 1000$ (dimensionless units) of the knallgas reaction in a 1D chamber. Black: hydrogen ($\tilde{c}_1$), red: oxygen ($\tilde{c}_2$), green: oxyhydrogen ($\tilde{c}_3$).

5.1.4 Heat equation

The knallgas reaction is exothermic and heats up the reaction chamber. The temperature evolution is given by the heat equation (4.1) (cross diffusion neglected) but now with an additional source term coming from dc_i/dt in equation (4.36),

$$\partial_t T = \kappa \nabla^2 T + \frac{q(\tilde{c}_i)}{c_v \rho} . \tag{5.16}$$

Note that the source term is directly related to the entropy production rate (5.11),

$$q(\tilde{c}_i) = K \tilde{c}_1^2 \tilde{c}_2 \left[M_1(2\hat{\mu}_1 - 2\hat{\mu}_3) + M_2(\hat{\mu}_2 - \hat{\mu}_3) \right] = T\,\sigma . \tag{5.17}$$

5.2 Self-organized chemical patterns

In 1951, Belousov[2] found persistent oscillations of cerium(IV) and cerium(III) ions in a certain chemical reaction. These oscillations were manifested by a periodic color change of the solution with a rather constant frequency. This was one of the first hints that chemical reactions must not end in a stationary equilibrium state but may also show rather involved temporal (and also spatial) behavior, including chaotic dynamics, a true evidence that nonlinear effects far from equilibrium are crucial.

Later on, the fairly complicated reaction scheme of the nowadays called Belousov–Zhabotinsky reaction was simplified by a kinetic model, now known as *Oregonator*, after its birthplace Oregon, USA. An even simpler model, showing also chemical oscillations, was derived in the 60ies and 70ies by Prigogine[3] and Levefer at the Free University of Brussels, Belgium and named thereafter as *Brusselator*. The Brusselator equations and their extensions, including spatial diffusion terms, allow for the description of chemical instabilities, oscillations, and pattern formation [6].

We shall discuss this model for the rest of the chapter.

5.2.1 The Brusselator

To describe persistent chemical oscillations, we need more than one reaction and a constant feed and removal of the consumed reactants and the products, respectively. Under these circumstances, we are dealing with an open system and entropy is produced as long as the reactions are kept alive. In this way, the emergence of temporal or spatial order does not contradict to the second law.

Instead of a real chemical reaction scheme the Brusselator is based on four model reactions with six reactants that are not further specified. The model reactions read

$$\mathrm{A} \xrightarrow{K_1} \mathrm{X}, \tag{5.18a}$$

2 Boris P. Belousov, Russian chemist, 1893–1970.

3 Ilya Prigogine, Russian-Belgian physical chemist, 1917–2003.

$$\mathrm{B} + \mathrm{X} \xrightarrow{K_2} \mathrm{Y} + \mathrm{C}, \tag{5.18b}$$

$$2\mathrm{X} + \mathrm{Y} \xrightarrow{K_3} 3\mathrm{X}, \tag{5.18c}$$

$$\mathrm{X} \xrightarrow{K_4} \mathrm{E}, \tag{5.18d}$$

where K_i are the reaction rates, and, for simplicity, backwards reactions are neglected. Reaction (5.18c) is autocatalytic and accounts for the nonlinearities in the kinetic equations. From the stoichiometry, it follows that all molar masses must be equal, $M_i = M$. We assign the molar densities

$$\mathrm{X} \to \tilde{c}_1, \quad \mathrm{Y} \to \tilde{c}_2, \quad \mathrm{A} \to \tilde{c}_3, \quad \mathrm{B} \to \tilde{c}_4, \quad \mathrm{C} \to \tilde{c}_c, \quad \mathrm{E} \to \tilde{c}_e .$$

To obtain non-dimensional variables, we scale the molar densities with a certain reference molar density (e. g. ρ/M),

$$c_i = \frac{M}{\rho}\,\tilde{c}_i$$

as already done for equation (2.87). First we neglect again diffusion and obtain from (5.18) the rate equations

$$\frac{dc_1}{dt} = k_1 c_3 - k_2 c_4 c_1 + k_3 c_1^2 c_2 - k_4 c_1, \tag{5.19a}$$

$$\frac{dc_2}{dt} = k_2 c_4 c_1 - k_3 c_1^2 c_2, \tag{5.19b}$$

$$\frac{dc_3}{dt} = -k_1 c_3 + S_3, \tag{5.19c}$$

$$\frac{dc_4}{dt} = -k_2 c_1 c_4 + S_4 . \tag{5.19d}$$

The reaction rates k_i differ from K_i in (5.18) by powers of the reference density due to scaling of $\tilde{c}_i$ and all have the unit 1/s. The two source terms S_3, S_4 allow for feeding the reactor with the consumed reactants A, B. Thus we have an open system, and the reactions must not come to an end. If the sources are adjusted in a convenient matter, we may assume c_3, c_4 constant, and only two nonlinearly coupled ordinary differential equations (5.19a), (5.19b) survive. Three of the four reaction rates can be scaled into c_3, c_4 and the time and one obtains

$$\begin{aligned}\frac{dc_1}{dt} &= a - (b + k_0)c_1 + c_1^2 c_2, \\ \frac{dc_2}{dt} &= bc_1 - c_1^2 c_2,\end{aligned} \tag{5.20}$$

with the constants (control parameters)

$$a = \frac{k_1}{k_3} c_3, \quad b = \frac{k_2}{k_3} c_4, \quad k_0 = \frac{k_4}{k_3}$$

and a time scaled by k_3. Equations (5.20) (with $k_0 = 1$) constitute the classical rate equations of the Brusselator model without diffusion.

5.2.2 Chemical oscillations

A stationary non-equilibrium solution of (5.20) is

$$c_1^{(0)} = \frac{a}{k_0}, \quad c_2^{(0)} = \frac{bk_0}{a} . \tag{5.21}$$

If we consider small disturbances

$$c_1(t) = c_1^{(0)} + u_1(t), \quad c_2(t) = c_2^{(0)} + u_2(t)$$

and neglect all nonlinear terms in u_i, we find from (5.20), after eliminating u_2, the equation for a damped harmonic oscillator,

$$d_{tt}^2 u_1 + \gamma d_t u_1 + \frac{a^2}{k_0} u_1 = 0$$

with damping rate

$$\gamma = k_0 - b + \frac{a^2}{k_0^2} .$$

As long as $\gamma > 0$ all disturbances u_i are exponentially decaying in time and the state (5.21) is reconstructed, defining (5.21) as a stable solution. But when $\gamma < 0$ (negative damping), the disturbances will grow exponentially in time while oscillating with the frequency $a/\sqrt{k_0}$. These oscillations serve as an heuristic explanation of the chemical oscillations detected experimentally in the late 50ties by Belousov, and which finally motivated the derivation of the Brusselator. The instability condition is thus

$$b \geq b_c = k_0 + \frac{a^2}{k_0^2},$$

where b_c is called the critical b. However, if $b > b_c$, the amplitudes u_i do not stay small but grow fast, and the nonlinearities become important and lead to saturation. To solve the problem well above threshold, one can resort to numerical solutions of (5.20), Figure 5.2.

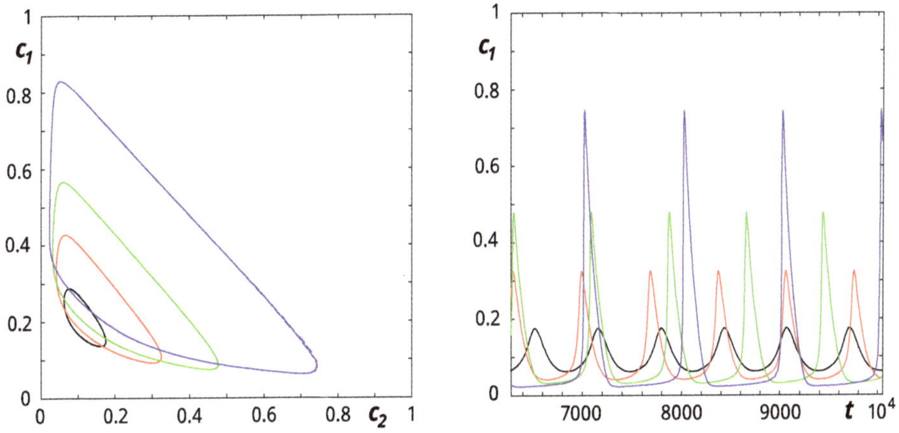

Figure 5.2: Numerical solutions of the Brusselator model without diffusion with $k_1 = k_2 = k_3$, $k_0 = 0.01$, $a = k_0^{3/2}$, $b_c = 2k_0$. Left: c_1 over c_2, limit cycles, right: c_1 over time, nonlinear oscillations. Values for b: black $b = 1.1b_c$, red $b = 1.4b_c$, green $b = 1.7b_c$, blue $b = 2.2b_c$. Time in dimensionless units of equation (5.20).

Knowing the temporal behavior of c_1, c_2 numerically, we are in the position to compute the entropy production

$$\sigma = -\sum_r^4 \omega_r \frac{A_r}{T} \tag{5.22}$$

with the reaction velocities from (5.18) according to

$$\omega_1 = k_1 c_3, \quad \omega_2 = k_2 c_1 c_4, \quad \omega_3 = k_3 c_1^2 c_2, \quad \omega_4 = k_4 c_1 .$$

Note that the stationary solution (5.21) does not mean all $\omega_i = 0$ (detailed balance), but rather $\omega_1 = \omega_4$ and $\omega_2 = \omega_3$. Then, σ is larger zero, even for $b < b_c$. To compute the affinities, we need the chemical potentials of all reactants in (5.18). If we assume the expressions for an ideal gas (2.87) we may write

$$\mu_j = g_j(p, T) + RT \ln c_i .$$

Then, $A_r = \sum_j \nu_{rj}\mu_j$ can be computed as

$$\begin{aligned}
A_1 &= RT(\ln c_1 - \ln a), \\
A_2 &= RT(\ln c_2 + \ln c_c - \ln b - \ln c_1), \\
A_3 &= RT(\ln c_1 - \ln c_2), \\
A_4 &= RT(\ln c_e - \ln c_1),
\end{aligned}$$

where c_c, c_e denote the molar densities of the products C and E, respectively, and all g_j are assumed being equal. To obtain negative affinities for all times, the products C, E should be depleted, and we may assume small concentrations, putting

$$\ln c_c = \ln c_e = -100 \, .$$

Figure 5.3 shows $\sigma(t)$ and the totally produced entropy

$$\Delta S(t) = \int_0^t dt' \; \sigma(t') \tag{5.23}$$

for the simulations and parameters of Figure 5.2.

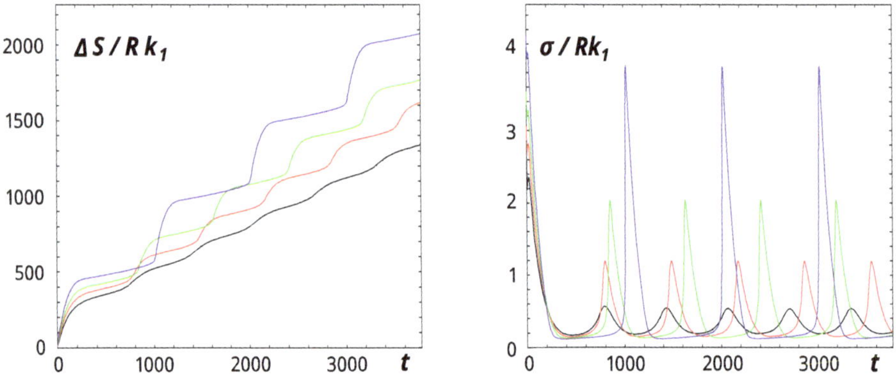

Figure 5.3: Entropy (left) and entropy production rate (right) for the solutions shown in Figure 5.2, $k_1 = k_2 = k_3$, $k_0 = 0.01$, $a = k_0^{3/2}$, black $b = 1.1b_c$, red $b = 1.4b_c$, green $b = 1.7b_c$, blue $b = 2.2b_c$.

5.2.3 Reaction-diffusion equations and spatial pattern formation

After the seminal work of Belousov on chemical oscillations, Anatoli Zhabotinsky continued the experimental work in the 1960ies and found spatial patterns in chemical non-equilibrium reactions, such as traveling wave fronts, spirals or targets. Again about 30 years later, stationary non-equilibrium regular spatial patterns predicted already in the 1950ies by Alan Turing[4] on theoretical grounds were observed by a French group in Bordeaux (De Kepper et al.), and, in the meantime, in Austin, Texas, by Swinney et al.

The simple Brusselator scheme (5.20) may show both temporally oscillating and spatially periodic structures if diffusion terms are added. Extending the treatment to spatial

4 Alan Turing, British mathematician, computer scientist, and theoretical biologist, 1912–1954.

patterns, we may assume $c_n = c_n(x, y, t)$ for the concentrations. We only consider the dependence on two (horizontal) spatial coordinates, because the experiments are usually performed in a shallow, and therefore plane and two-dimensional reaction chamber. Rescaling c_i, a, b, and t, one obtains $k_0 = 1$. Then the reaction-diffusion equations for the Brusselator read

$$\frac{\partial n_1}{\partial \tilde{t}} = \tilde{a} - (\tilde{b} + 1)n_1 + n_1^2 n_2 + D_1 \nabla^2 n_1, \tag{5.24a}$$

$$\frac{\partial n_2}{\partial \tilde{t}} = \tilde{b} n_1 - n_1^2 n_2 + D_2 \nabla^2 n_2 \,, \tag{5.24b}$$

where we neglect cross diffusion terms. Here, $n_i = c_i/\sqrt{k_0}$, $\tilde{a} = a/k_0$, $\tilde{b} = b/k_0$ and $\tilde{t} = t\, k_0$.

5.2.3.1 Linear analysis

Before presenting numerical results of the fully nonlinear equations, we perform a linear stability analysis around the still existing fixed point (5.21) (we leave the tildes at a, b, t). Substituting

$$n_1(x, t) = a + u_1\, \mathrm{e}^{ikx + \lambda t}, \quad n_2(x, t) = \frac{b}{a} + u_2\, \mathrm{e}^{ikx + \lambda t}$$

into (5.24), where k is the wave vector of the disturbances and taking only the linear terms, (5.24) reads

$$\lambda u_1 = (b - 1)u_1 + a^2 u_2 - D_1 k^2 u_1, \tag{5.25a}$$

$$\lambda u_2 = -b u_1 - a^2 u_2 - D_2 k^2 u_2 \,. \tag{5.25b}$$

The solvability condition yields the characteristic polynomial

$$\lambda^2 + p_1 \lambda + p_0 = 0 \tag{5.26}$$

with

$$p_1 = a^2 - b + 1 + (D_1 + D_2)k^2, \quad p_0 = a^2 + k^2((1 - b)D_2 + a^2 D_1) + D_1 D_2 k^4 \,.$$

A positive real part of at least one of the roots of (5.26) marks the onset of instability. There are two qualitatively different possibilities regarding how the fixed point (5.21) may become unstable:

(i) Monotonic mode, *Turing mode*, [7]. $\lambda = 0$ is a solution of (5.26) if $p_0 = 0$. If a is considered to be fixed and b is used as variable control parameter, $p_0 = 0$ can be solved for b:

$$b_c^T = \frac{a^2 + (D_1 a^2 + D_2)k^2 + D_1 D_2 k^4}{D_2 k^2} \,. \tag{5.27}$$

This is a function with a minimum at (see Figure 5.4)

$$k_c^2 = \frac{a}{\sqrt{D_1 D_2}}\,, \tag{5.28}$$

leading to

$$b^T = b_c^T(k_c) = (a\sqrt{\tilde{D}} + 1)^2 \tag{5.29}$$

with $\tilde{D} \equiv D_1/D_2$. Increasing b, a with respect to time monotonically growing instability sets in if b exceeds b^T; the wave number of this mode is given by (5.28).

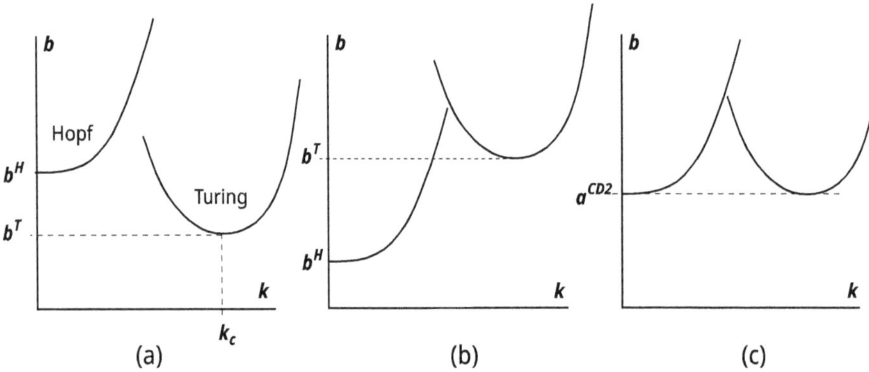

Figure 5.4: Sketch of Turing and Hopf instability for the Brusselator with diffusion. If increasing b and $a < a^{CD2}$, the Turing instability comes first (a), if $a > a^{CD2}$ the Hopf instability occurs above threshold (b). For (c) is $a = a^{CD2}$.

(ii) Oscillatory mode, *Hopf mode*, named after E. Hopf.[5] Here, λ has a non-vanishing imaginary part at threshold, and $\lambda = i\omega$ yields in (5.26)

$$p_1 = 0, \quad \omega^2 = p_0 > 0\,.$$

Thus

$$b_c^H = D\,k^2 + a^2 + 1 \tag{5.30}$$

with $D \equiv D_1 + D_2$ and

$$k_c = 0, \quad b^H = b_c^H(0) = a^2 + 1\,. \tag{5.31}$$

5 Eberhard Hopf, German mathematician, 1902–1983.

The situation where both instabilities set in at the same value of b (compare Figure 5.4) is a point in 2D-parameter space a, b, called codimension-two-point (CD2-point), determined from the condition

$$b^T(a) = b^H(a).$$

Solving for a this yields

$$a^{CD2} = \frac{2\sqrt{\tilde{D}}}{1-\tilde{D}}, \quad b^{CD2} = \left(\frac{1+\tilde{D}}{1-\tilde{D}}\right)^2. \tag{5.32}$$

The CD2-point can only exist for $\tilde{D} < 1$. Figure 5.5 shows a sketch of the parameter plane for $\tilde{D} < 1$. Depending on a, both instability types are possible.

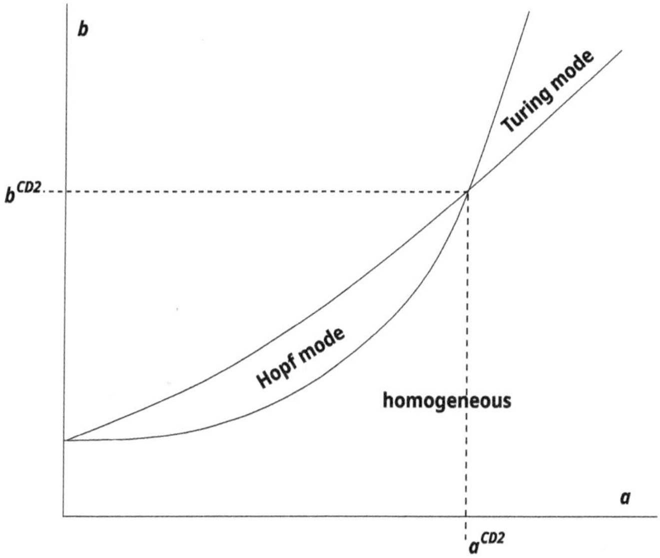

Figure 5.5: Sketch of 2D-parameter plane for the Brusselator with $\tilde{D} < 1$. On the left-hand side of the dashed line the homogeneous state becomes oscillatory unstable; on the right-hand side, one observes Turing patterns.

5.2.3.2 Nonlinear analysis

Next we wish to integrate numerically the fully nonlinear system (5.24) applying a FD scheme[6] as already done in Section 4.1.2. We begin with $D_1 = 0.1$, $D_2 = 1$. For a above the codimension-two-point $a^{CD2} \approx 0.7$, we obtain Turing patterns normally in the form of more or less regular hexagons, including a large number of point and line defects that

6 A more advanced Matlab code for the numerical integration can be found in [5].

originate from the randomly disturbed initial conditions

$$n_1(x,y,0) = a\,(1 + A\xi_1(x,y)), \quad n_2(x,y,0) = \frac{b}{a}\,(1 + A\xi_2(x,y))$$

with $\xi_i(x,y)$ equally distributed in $[-0.5, 0.5]$ and $A = 0.1$. A time series is shown in Figure 5.6.

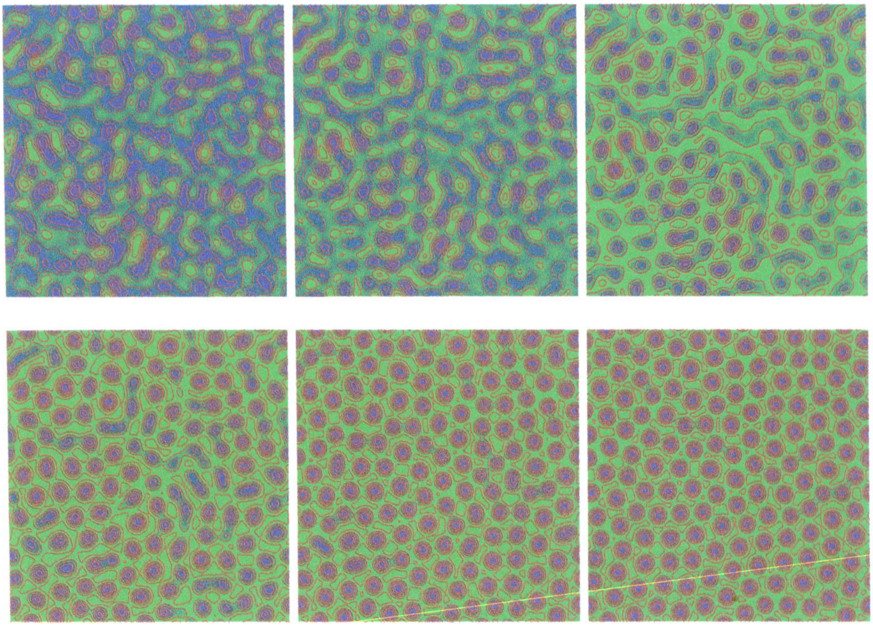

Figure 5.6: Numerical solutions of the fully nonlinear model in the Turing region; time series from top left to bottom right, grid size 300 × 300. Starting with a randomly disturbed pattern, for $a = 1$ asymptotically stationary Turing structures are found ($b = 3.5$).

Below the CD2-point, large-scaled regions in space oscillating with about the Hopf frequency can be seen shown in Figure 5.7 and in agreement with the linear theory.

Qualitatively different structures, namely the experimentally known target, fronts or spiral patterns, are found for other diffusion constants, for instance $D_1 = 1.5$, $D_2 = 1$, Figure 5.8. Note that for $D_1 > D_2$ the CD2-point is not defined, and the Turing region does not exist anymore.

Although the diffusive Brusselator is a very simple and constructed model, which has not much in common with the reactions studied by Belousov and Zhabotinsky, it can serve as a system, where non-equilibrium spatial and temporal patterns are created by self-organization. For the reader who wants to delve in deeper in the theory of self-organized pattern formation, we refer to the excellent book of H. Haken and references therein [6].

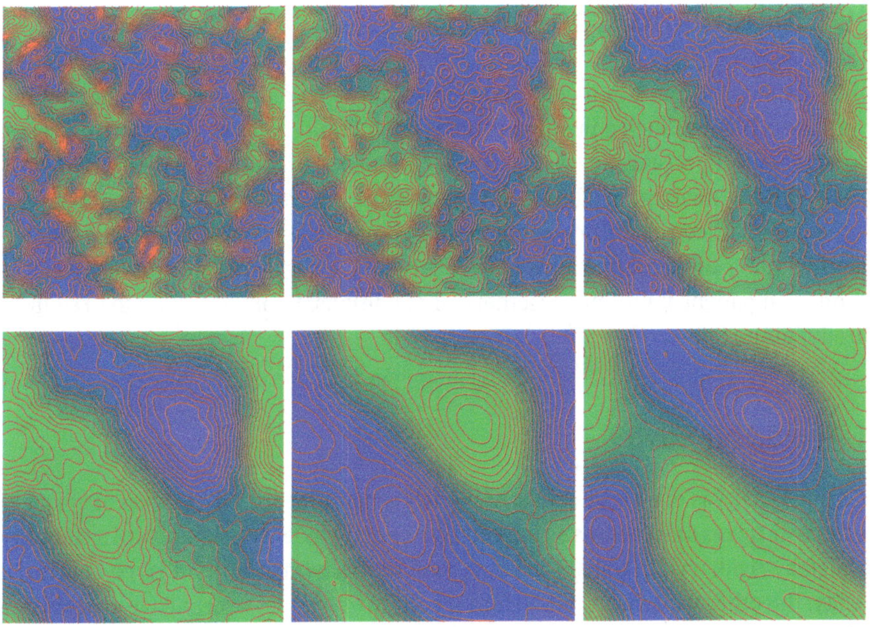

Figure 5.7: Numerical solutions of the nonlinear model in the Hopf region. For the value of $a = 0.5$, spatially large-scaled and temporally oscillating patterns emerge.

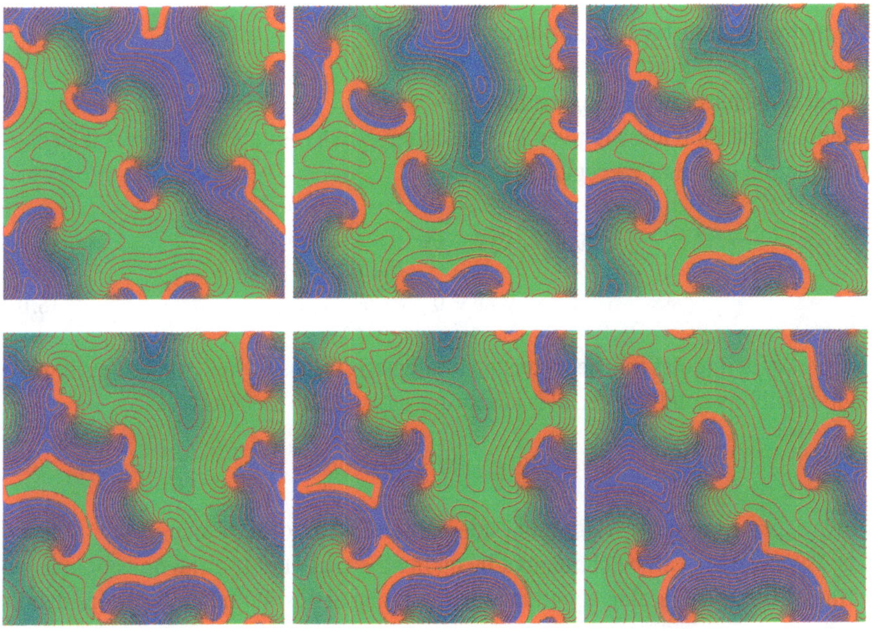

Figure 5.8: Spreading fronts are observed for other parameters in the oscillatory region. They resemble the chemical non-equilibrium patterns observed experimentally in the Belousov–Zhabotinsky reaction. Time series for $\tilde{D} = 3/2$, $a = 1$, $b = 3.5$.

6 Diffusion and transport

In this chapter, we consider a continuous medium with volume elements moving relatively to each other. Such a medium is normally a liquid, a gas or, in more rare cases, an elastic solid. In a realistic system, irreversible friction and dissipation are always present and the medium produces entropy at least as long as relative motion exists. The medium can consist of one component or a mixture of several miscible components, including chemical reactions between them. We start with the most simple case of a one-component fluid under isothermal conditions, finally entering the realm of hydrodynamics [8].

6.1 The driven cavity

The driven cavity is one of the standard models accounting for the creation of vorticies due to shear forces and eventually the emergence of turbulence. It consists of a chamber confined by rigid, no-slip walls. The upper wall moves with a constant given tangential velocity; the other walls are fixed. Before we study the model in detail, we concentrate on the equations that describe the spatio-temporal evolution of the velocity field inside the chamber.

6.1.1 Hydrodynamic basic equations

We shall consider only incompressible fluids with $\rho = \rho_0 =$ constant. Conservation of momentum and total mass is given by the Navier–Stokes equations (3.4) and the continuity equation (3.2) that we repeat here:

$$\partial_t \vec{v} + (\vec{v} \cdot \nabla)\, \vec{v} = \nu \nabla^2 \vec{v} - \frac{1}{\rho_0} \nabla P\,, \qquad (6.1a)$$

$$\nabla \cdot \vec{v} = 0\,, \qquad (6.1b)$$

where $\nu = \eta/\rho_0$ denotes the, here constant, kinematic viscosity. In three dimensions the incompressibility condition (6.1b) is identically satisfied by the ansatz

$$\vec{v} = \nabla \times (\Psi \hat{e}_z) + \nabla \times \left[\nabla \times (\Phi \hat{e}_z)\right]. \qquad (6.2)$$

The two expressions on the right-hand side are denoted as toroidal and poloidal velocity fields. The pressure can be eliminated by taking the curl of (6.1a), and one obtains two nonlinearly coupled, scalar equations for the scalar fields Ψ and Φ (for more details see [5]).

https://doi.org/10.1515/9783110794137-007

6.1.2 Plane flows: stream function and vorticity

Things are much easier if we restrict ourselves to the two-dimensional case; we do starting here. For a plane flow (6.2) simplifies to

$$\vec{v} = \nabla \times (\Psi \hat{e}_z) = \begin{pmatrix} \partial_y \Psi \\ -\partial_x \Psi \end{pmatrix} . \tag{6.3}$$

The field $\Psi(x, y, t)$ is called *stream function*; the curves with $\Psi(x, y, t) = \text{const.}$ are the stream lines. Due to ($t = \text{const.}$)

$$d\Psi = \partial_x \Psi \, dx + \partial_y \Psi \, dy = -v_y dx + v_x dy$$

one finds along a stream line, where $d\Psi = 0$:

$$\frac{dy}{dx} = \frac{v_y}{v_x} .$$

Hence, for the stationary case $\partial_t \Psi = 0$, the stream lines are identical with the particle paths.[1]

The vorticity of a given flow is defined by the curl of the velocity field

$$\vec{\xi} = \nabla \times \vec{v} .$$

In two dimensions and with (6.3), this simplifies to the scalar field (z-component of $\vec{\xi}$)

$$\xi = -\nabla^2 \Psi . \tag{6.4}$$

Taking the curl of (6.1a) one arrives at an evolution equation for ξ having the form of a transport equation:

$$\partial_t \xi + v_x \partial_x \xi + v_y \partial_y \xi = \nu \nabla^2 \xi . \tag{6.5}$$

This is a PDE for ξ that has to be completed with boundary and initial conditions. Moreover it contains the velocity field $\vec{v}$, which must be determined via (6.3) from Ψ by inverting (6.4).

6.1.3 Boundary conditions

Now we turn to the driven cavity. Figure 6.1 shows the system and the boundary conditions. We assume no-slip boundary conditions, on a rigid wall, the fluid adheres to the

1 In fluid dynamics, the notion "particle" stands rather for a volume element of the continuum swimming with the flow.

wall. In particular at the upper wall one has $v_x = U_0$, $v_y = 0$ or

$$\partial_x \Psi = 0, \quad \partial_y \Psi = U_0 .$$

On all other walls $\vec{v} = 0$ demands for

$$\hat{t} \cdot \nabla \Psi = 0, \quad \textit{and} \quad \hat{n} \cdot \nabla \Psi = 0 \tag{6.6}$$

with $\hat{n}$ perpendicular and $\hat{t}$ parallel to the wall.

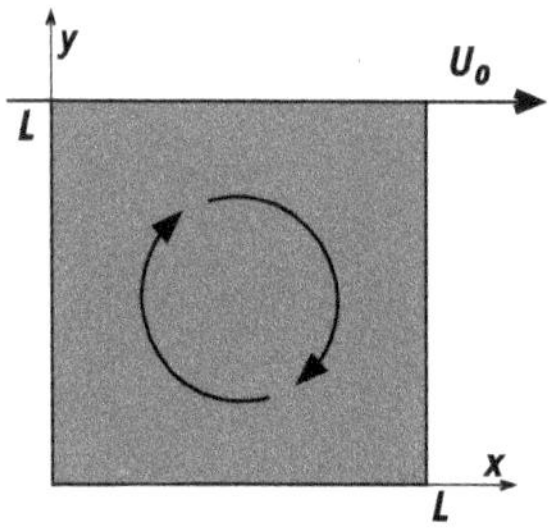

Figure 6.1: Sketch of the driven cavity. The upper wall moves with the constant velocity U_0 to the right. Inside the fluid (gray) vorticities are created.

Boundary conditions are necessary in (6.5) and for the inversion of (6.4). We take for example, the wall at $x = 0$. From $\vec{v} = 0$, it follows from (6.3)

$$\partial_x \Psi|_{x=0} = \partial_y \Psi|_{x=0} = 0 .$$

Hence, Ψ must be constant along the wall, and since only derivatives of Ψ occur, we may put

$$\Psi|_{x=0} = 0, \quad \partial_x \Psi|_{x=0} = 0 . \tag{6.7}$$

But now we have two conditions for Ψ, namely Dirichlet and Neumann, but none for the vorticity ξ. To end up with a well-defined problem, one must somehow "distribute" the conditions to the two fields Ψ and ξ. Equation (6.4) can be uniquely inverted with the Dirichlet condition $\Psi = 0$. To solve (6.5), we must find a condition for the vorticity ξ by the help of the Neumann condition. To this end, we evaluate (6.4) at the wall $x = 0$:

$$\partial_{xx} \Psi|_{x=0} = -\xi(0, y), \tag{6.8}$$

where we used $\partial_{yy} \Psi = 0$ due to (6.7). Expanding Ψ close to the wall yields

$$\Psi(\epsilon, y) = \Psi(0, y) + \epsilon \, \partial_x \Psi|_{x=0} + \frac{1}{2} \epsilon^2 \partial_{xx} \Psi \Big|_{x=0} + O(\epsilon^3) .$$

Inserting this into (6.8), we obtain

$$\xi(0,y) = -\lim_{\epsilon\to 0}\left[\frac{2}{\epsilon^2}\left(\Psi(\epsilon,y) - \Psi(0,y) - \epsilon\,\partial_x\Psi|_{x=0}\right)\right],$$

a Dirichlet boundary condition for ξ at $x = 0$. If we apply (6.7), this reduces to

$$\xi(0,y) = -\lim_{\epsilon\to 0}\frac{2\Psi(\epsilon,y)}{\epsilon^2}\,. \tag{6.9}$$

Numerically, one evaluates the r. h. s. of (6.9) by putting $\epsilon = \Delta x$, where Δx denotes the step size of a finite difference method (for details we refer to [5]). The same evaluation can be done on the other walls and yields

$$\xi(L,y) = -\lim_{\epsilon\to 0}\frac{2\Psi(L-\epsilon,y)}{\epsilon^2}, \tag{6.10a}$$

$$\xi(x,0) = -\lim_{\epsilon\to 0}\frac{2\Psi(x,\epsilon)}{\epsilon^2}\,. \tag{6.10b}$$

On the top wall, we have $\partial_y\Psi = U_0$, and the boundary condition for ξ reads

$$\xi(x,L) = -\lim_{\epsilon\to 0}\left[\frac{2\Psi(x,L-\epsilon)}{\epsilon^2} - \frac{2U_0}{\epsilon}\right]. \tag{6.11}$$

6.1.4 Dimensionless equations and Reynolds number

In fluid mechanics one normally works with dimensionless variables and equations. For our problem, the scaling

$$\vec{v} = U_0\,\tilde{\vec{v}}, \quad (x,y) = (L\tilde{x}, L\tilde{y}), \quad t = \frac{L}{U_0}\,\tilde{t}$$

is appropriate and leads instead of (6.4), (6.5) to the system (all tildes are again removed)

$$\partial_t\xi + v_x\partial_x\xi + v_y\partial_y\xi = \frac{1}{R_e}\nabla^2\xi, \tag{6.12a}$$

$$\xi = -\nabla^2\Psi \tag{6.12b}$$

with the *Reynolds number*[2]

$$R_e = \frac{U_0 L}{\nu}$$

as the only parameter left (characteristic number).

2 Osborne Reynolds, British physicist, 1842–1912.

6.1.5 Results

We apply a finite difference method to express the spatial derivatives of (6.12). Equation (6.12b) is inverted by a standard method; the time integration of (6.12a) is achieved by a semi-implicit Crank–Nicolson scheme; for details see [5, 9].

We present results on a 200 × 200 grid with $\Delta x = 1/199$ and for different Reynolds numbers. The stream lines at several times are depicted in Figures 6.2–6.5. To get a perspective of the time and length scales, we go back to dimensionful quantities and take water as the work fluid with $\nu = 10^{-6}\mathrm{m}^2/\mathrm{s}$. For $R_e = 1000$ and $L = 0.1\,\mathrm{m}$ a lid speed of $U_0 = 1\,\mathrm{cm/s}$ results. For the time scale, one finds $\tau = L/U_0 = 2\,\mathrm{s}$. At $R_e = 30000$ (Figure 6.5) and the same cavity size, one has $U_0 = 0.3\,\mathrm{m/s}$.

For $R_e = 100$, corresponding to a very slow lid speed of 1 mm/s, a stationary state is reached after $t \approx 2000$ s, Figure 6.2. For larger R_e, more complicated patterns, but after a certain relaxation time still stationary patterns occur, Figure 6.3.

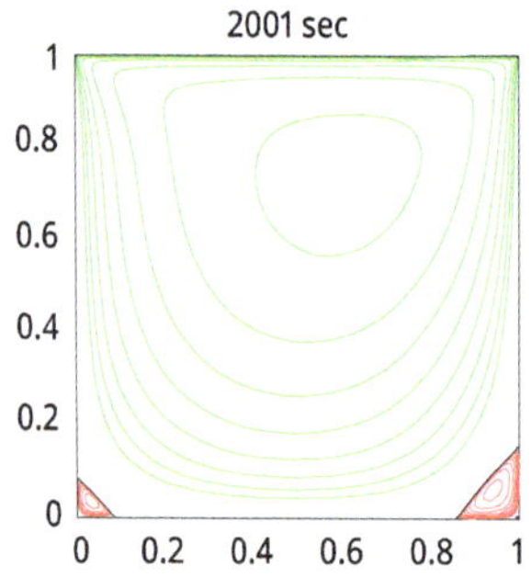

Figure 6.2: Driven cavity; stationary solution for $R_e = 100$ reached after about 2000 seconds. (water, cavity length 10 cm, $U_0 = 1\,\mathrm{mm/s}$).

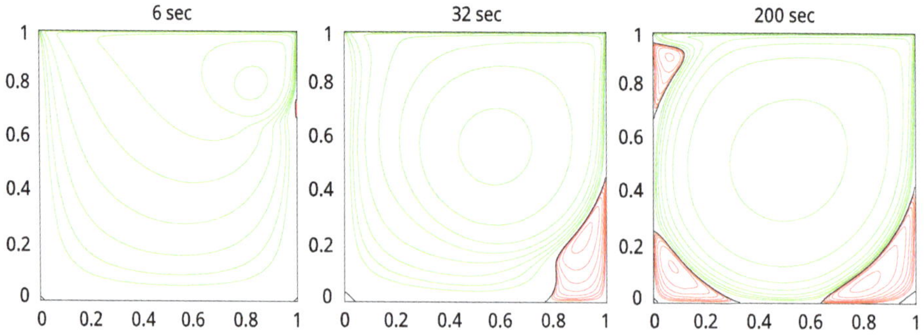

Figure 6.3: Time series for $R_e = 5000$. The state in the right frame is stationary. Times are for water, cavity length 10 cm, $U_0 = 5\,\mathrm{cm/s}$.

As it is known from the literature [10], the stationary vorticies for smaller R_e become oscillatory unstable at about $R_e \approx 8000$, which can be seen from the time series in Figures 6.4, 6.5.

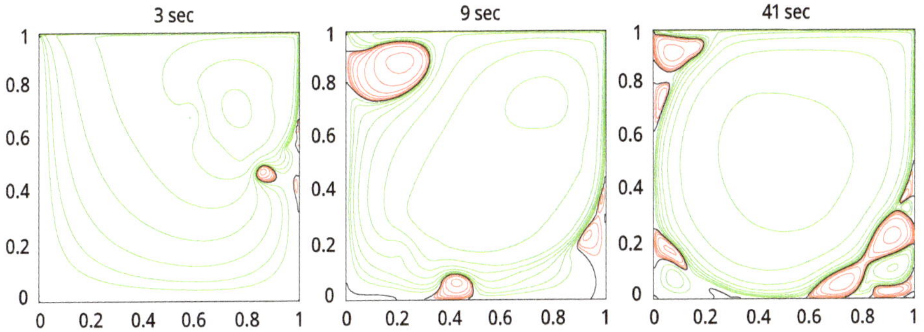

Figure 6.4: Time series for $R_e = 20000$. Here the stream lines remain time dependent, $U_0 = 20\,\mathrm{cm/s}$.

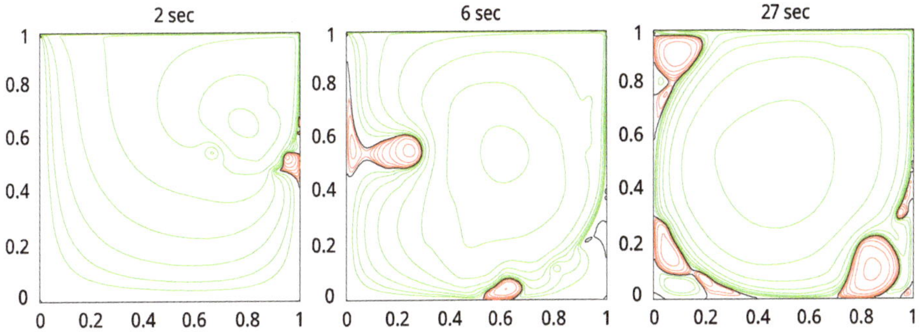

Figure 6.5: Time series for $R_e = 30000$. The spatio-temporal behavior of the fluid patterns gets more and more chaotic, $U_0 = 30\,\mathrm{cm/s}$.

6.1.6 Entropy production

From the velocity field computed in the last section, we can find the entropy production. For an isothermal incompressible flow, equation (3.12) with (3.13) applies. Expressing the velocity by the stream function yields

$$\tilde{\sigma}(x,y,t) = 2\left(\frac{\partial^2\Psi}{\partial x\partial y}\right)^2 + \frac{1}{2}\left(\frac{\partial^2\Psi}{\partial x^2} - \frac{\partial^2\Psi}{\partial y^2}\right)^2 , \tag{6.13}$$

where $\tilde{\sigma}$ is the dimensionless production rate, which should be scaled with

$$\sigma = \frac{2\rho_0 \nu^3 R_e^2}{TL^4}\,\tilde{\sigma} .$$

The total production rate is given from (6.13) by integrating over x, y as

$$\tilde{\Sigma}(t) = \int_0^1 dx \int_0^1 dy\, \tilde{\sigma}(x,y,t) . \tag{6.14}$$

If the Navier–Stokes equations would be linear, a unique solution would exist that would simply scale linearly with U_0. In the dimensionless variables we use, the velocities are already scaled with U_0, and therefore $\tilde{\sigma}$ would not depend on U_0 and R_e. Looking at Figure 6.6, this is obviously the case for very small U_0 and $\tilde{\sigma}$ approaches $R_e = 0$ with zero slope. But already for rather small values of R_e, nonlinearities become effective, leading to the formation of extra vorticies in the cavity, and finally to a stronger and stronger increase of the entropy production.

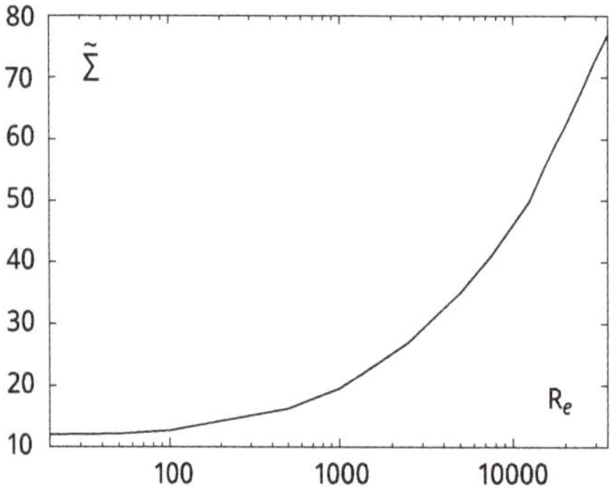

Figure 6.6: Dimensionless total entropy production rate (6.14) of the driven cavity over Reynolds number. For very small R_e, the rate becomes independent on R_e.

If R_e becomes larger, the structures stay time dependent, and vorticies are continuously formed, annihilated, and transported by the flow. This is expressed by a time dependent entropy production rate, as shown for various large R_e in Figure 6.7. The more complicated and chaotic pattern formation becomes, the higher are the fluctuations of $\tilde{\Sigma}$.

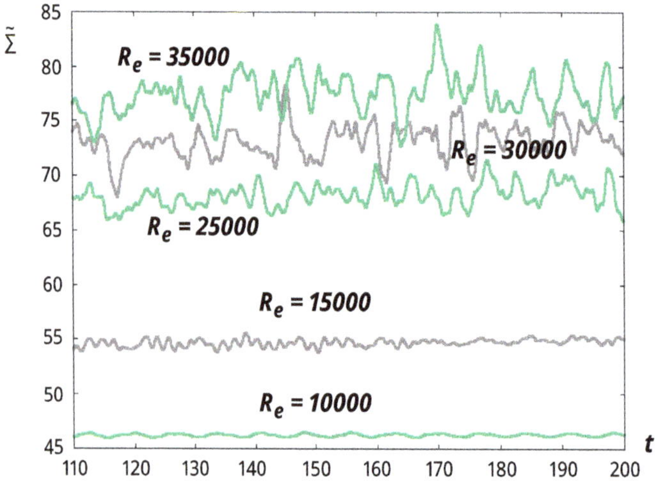

Figure 6.7: Total entropy production rate over time for larger R_e and time dependent structures.

6.2 Convection of a pure gas

Now we come back to the question of the temperature distribution in a heated room with insulating walls, already studied in Section 4.1.3. There we assumed pure thermal diffusion caused by heat conduction of the air in rest and found that it would take many days to obtain a somewhat homogeneous temperature across the room. Now we wish to include the motion of the air and extend the system by the Navier–Stokes equations. Contrary to the former system from Section 4.1.3, we simplify the situation and assume insulating walls on the left, right, and on the ceiling. Along the lower side, we prescribe a fixed and constant temperature T_0, corresponding to a floor heating. Window and stove are removed. The temporal behavior of the system is different to the open window problem in so far that for $t \to \infty$ it reaches thermal equilibrium with temperature T_0 everywhere and the air comes to rest again.

6.2.1 Basic equations and boundary conditions

Figure 6.8 sketches the system and the boundary conditions. For the air remaining in rest, the heat equation with the initial condition

$$T(x, y, t = 0) = T_1$$

can be solved analytically:[3]

$$T(x, y, t) = T_0 - \delta T \,\mathrm{erf}(y/(2\sqrt{\kappa t})) \tag{6.15}$$

with

$$\delta T = T_0 - T_1 \,.$$

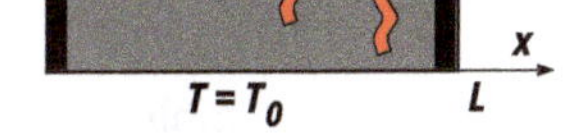

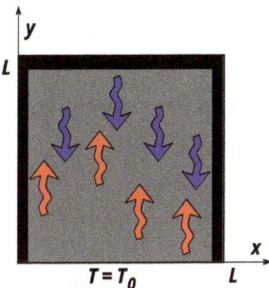

Figure 6.8: A room uniformly heated from below with perfectly insulating walls and ceiling.

3 The boundary condition at $y = L$ is thereby only approximately fulfilled, however the error remains exponentially small as long as $t \ll \tau$.

Here, κ again denotes the thermal diffusivity of the air, and $\mathrm{erf}(x)$ stands for the error function. The solution (6.15) describes a temperature front moving slowly from the floor to the ceiling by thermal diffusion. The time needed to cross the room is in the order of the thermal diffusion time

$$\tau = L^2/\kappa \tag{6.16}$$

and amounts for $L = 3\,\mathrm{m}$ about 5 days, as already computed in Section 4.1.3. To account for the effect of the air flow on the temperature distribution, the full energy equation (3.8) should be considered. However, we shall assume an incompressible gas ($\nabla \cdot \vec{v} = 0$) and neglect the heat produced by friction. For a gas, the assumption of incompressibility is not a good one; anyway we can use it here, because the main mechanisms of heat transport and buoyancy-driven instability are still present. With these approximations, we arrive at the heat equation of the form

$$\partial_t T + \vec{v} \cdot \nabla T = \kappa \nabla^2 T \, .$$

The air begins to move due to buoyancy. Warm air is lighter and tends to move upwards; cold air descends in a vertical gravitation field. Assuming a linear relation (Boussinesq approximation) between density ρ and temperature

$$\rho(T) = \rho_0(1 - \alpha(T - T_0)),$$

where

$$\alpha = \frac{1}{V}\frac{dV}{dT}\bigg|_{T=T_0} > 0$$

is the thermal expansion coefficient. The Navier–Stokes equations (3.3) have to be extended by an additional volume force density term and read

$$\frac{d\vec{v}}{dt} = \nu\nabla^2\vec{v} - \frac{1}{\rho_0}\nabla p - g(1 - \alpha(T - T_0))\hat{e}_y$$

with $\nu = \eta/\rho_0$ and the gravitation acceleration g. Since no reference velocity U_0 exists, the time scaling we choose here differs from that of Section 6.1.4:

$$\vec{v} = \frac{L}{\tau}\vec{\tilde{v}}, \quad (x, y) = (L\tilde{x}, L\tilde{y}), \quad t = \tau\,\tilde{t}, \quad T = T_0\,\Theta$$

with τ from (6.16). Taking again the representation (6.3) for a 2D-flow, the complete system thus reads (tildes removed)

$$\frac{1}{P_r}[\partial_t\xi + v_x\partial_x\xi + v_y\partial_y\xi] = \nabla^2\xi + R\,\partial_x\Theta, \tag{6.17a}$$

$$\xi = -\nabla^2\Psi, \tag{6.17b}$$

$$\partial_t\Theta + \vec{v}\cdot\nabla\Theta = \nabla^2\Theta \tag{6.17c}$$

with the *Prandtl number*[4] $P_r = \nu/\kappa$ and the *Rayleigh number*[5]

$$R = \frac{g\alpha L^3 T_0}{\kappa\nu} . \tag{6.18}$$

The system (6.17) has to be completed by boundary conditions. Again, no-slip conditions for the velocity, as in Section 6.1.3 ($U_0 = 0$), are assumed along all the boundaries. The ceiling and walls are perfect thermal insulaters,

$$\hat{n}\ \nabla\Theta = 0$$

with $\hat{n}$ perpendicular to the boundary. On the bottom, we put

$$\Theta = 1 ;$$

for the initial condition we use

$$\Theta = \frac{T_1}{T_0}, \quad \Psi = \xi = 0 .$$

6.2.2 Numerical results

The method applied in Section 6.1.5 is easily extended to the non-isothermal case; for details see [5]. The heat equation (6.17c) can be iterated fully explicitly by an FTCS method, as outlined in Section 4.1.2. This yields an upper bound for the time step [5]

$$\Delta t < \frac{1}{4}\Delta x^2 .$$

To obtain more accurate results, Δt should be much smaller. If we compute R for the values of air (see Appendix A.2.1, $\alpha = 1/273$ K as for an ideal gas), one finds for $L = 3$ m and $T_0 = 300$ K a huge value of $R \approx 10^{12}$. The air flow becomes turbulent almost immediately, and small-scaled vorticies are created, which demand for an extremely high resolution in space. Since we are not interested here in high-performance computing but only in qualitative results, we take the much smaller value $R = 10^9$ for which a 200×200 grid

$$\Delta x = 1/199, \quad \Delta t = 0.0003\Delta x^2 \approx 0.75 \cdot 10^{-8}$$

---

4 Ludwig Prandtl, German physicist, 1875–1953.

5 3rd Baron Rayleigh (John W. Strutt), British physicist, 1842–1919.

proves to be sufficient. As initial condition, we take

$$\Theta(x,y,0) = \begin{cases} 0.9 & \text{for } 0 < x \leq 1, \\ 1.0 & \text{for } x = 0\,. \end{cases}$$

The initial room temperature is thus everywhere $T_1 = 0.9\ T_0$.

Figure 6.9 depicts contour lines of the temperature field at several times. The first frame shows the diffusion front (6.15), which becomes unstable after about 4 minutes, and a vortex flow occurs, which entrains and quickly distributes the temperature field in the whole room. Thereby, heat is transported from the floor to the ceiling, and as a consequence thermalization become much more effective, which can be clearly seen at

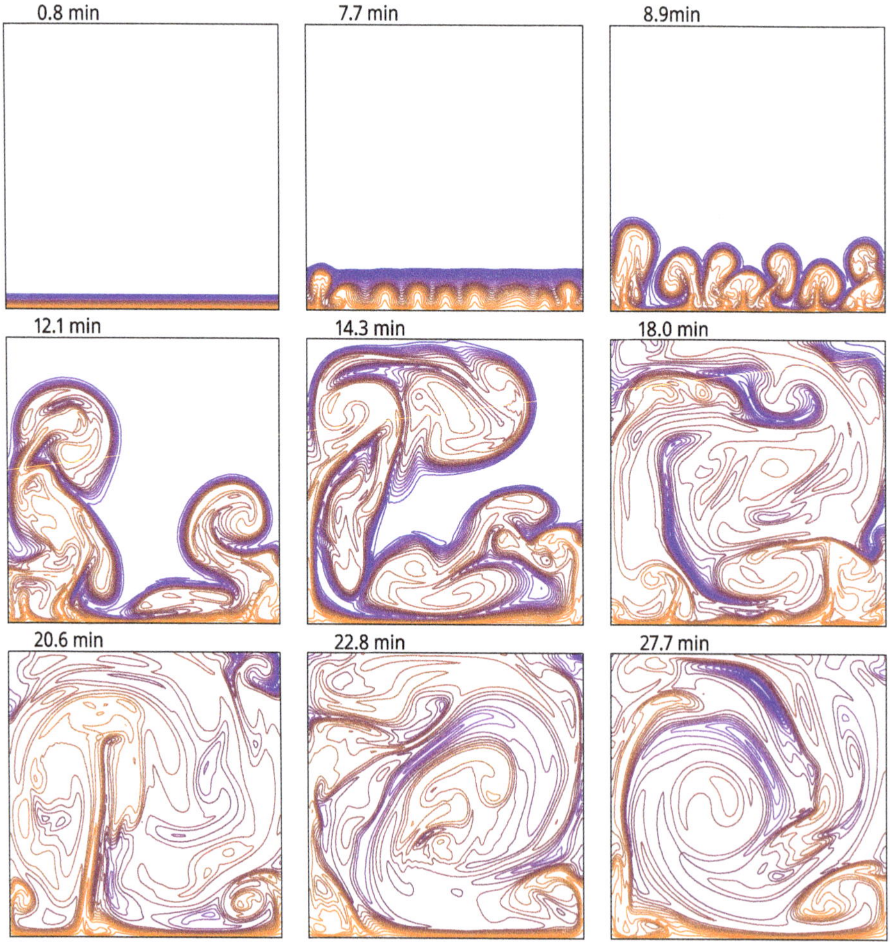

Figure 6.9: Evolution of the temperature field; time series for $R = 10^9$. The flow becomes turbulent soon. The times stated are valid for air ($\tau = 0.45 \cdot 10^6$ s).

the time dependence of the mean ceiling temperature given as

$$\Theta_1(t) = \int_0^1 dx\, \Theta(x, y = 1, t), \tag{6.19}$$

shown in Figure 6.10.

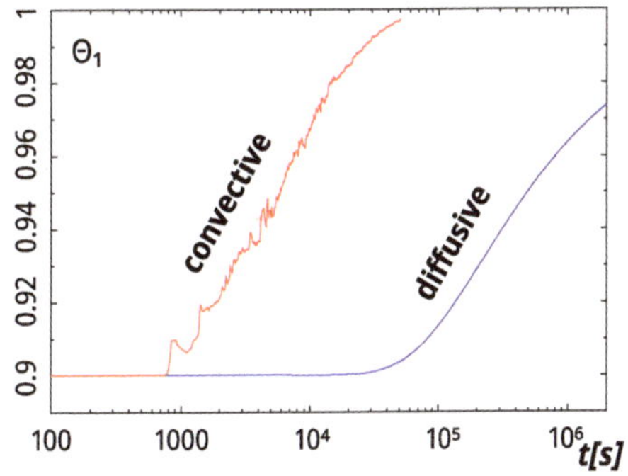

Figure 6.10: Averaged temperature of the ceiling (6.19) for convection ($R = 10^9$) and pure diffusion ($R = 0$). For this Rayleigh number, convection leads to an at least hundred times larger heat transport than diffusion alone.

6.2.3 Entropy production

As already mentioned, the whole process is transient and comes to an end in the long time limit $t \to \infty$ in thermal equilibrium with temperature T_0 everywhere. Then, no buoyancy exists anymore, and the flow has come to rest due to viscous damping. But if we consider the initial state and replace the heated floor for a moment with a wall with temperature T_1, this is also an equilibrium state. Thus we can compute the entropy difference of initial and final state using our equilibrium formulas derived in Chapter 1. Let us consider the initial state where the air has everywhere the temperature T_1 coming into sudden contact with a heat bath (the floor) with T_0. If we consider the heat bath and room as a closed system, equation (1.15) tells us the entropy increase

$$\Delta S/mc = \frac{T_1}{T_0} - 1 - \ln \frac{T_1}{T_0} \approx 0.00536 \tag{6.20}$$

for $T_1/T_0 = 0.9$. Now we can compute the entropy production rate according to equation (4.13). For a gas moving so slowly the entropy produced due to friction is completely negligible. Thus we find for the total entropy production

$$\langle \sigma(t) \rangle = \lambda \int_V dV \left(\frac{\nabla T}{T} \right)^2 = \lambda L \int_{\tilde{V}} d\tilde{V} \left(\frac{\nabla \Theta}{\Theta} \right)^2, \tag{6.21}$$

where tildes denote non-dimensional variables. The integral can easily be computed from the numerical solutions; the result is shown in Figure 6.11, left frame. From $\langle \sigma(t) \rangle$, the totally produced entropy at time t reads

$$\Delta S(t) = \int_0^t dt' \, \langle \sigma(t') \rangle = \frac{L^2}{\kappa} \int_0^{\tilde{t}} d\tilde{t}' \, \langle \sigma(\tilde{t}') \rangle = mc \int_0^t \int_{\tilde{V}} d\tilde{t}' d\tilde{V} \left(\frac{\nabla \Theta}{\Theta} \right)^2 . \tag{6.22}$$

Clearly, the produced entropy increases monotonically (Figure 6.11, right) and reaches asymptotically a value of $\approx$ 0.0058, less then 10 percents too large compared to (6.20), probably due to numerical errors and finite resolution.

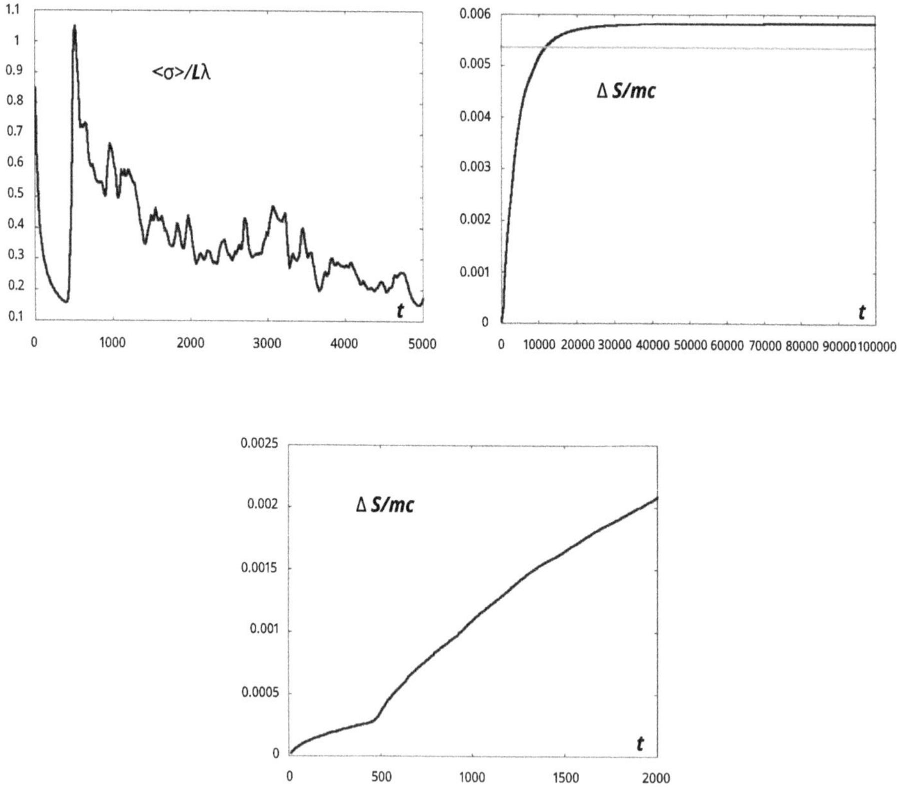

Figure 6.11: Left: integrated entropy production (6.21) inside the room; right: totally produced entropy, green line marks the value found from equilibrium thermodynamics (6.20); bottom: section of the right frame. The onset of convective patterns can be clearly detected by the sudden change of the slope.

The slope of the entropy becomes much larger if convective pattern formation sets in at about $t \approx 500$ s (Figure 6.11 bottom). There is often the argument that pattern formation contradicts the second law, because the creation of (ordered) structures should lower the entropy. Here, we see that the contrary is right: if patterns are formed, even more entropy is produced per time, because heat conduction (and friction) is increasing.

6.2.4 Linear stability analysis

It is difficult to compute the stability of the time and space-dependent diffusion front (6.15); we shall postpone this to the next paragraph. First we treat the simpler situation of a time independent base state and modify the top boundary conditions into

$$T(x,y=L) = T_1 \,. \tag{6.23}$$

A solution of (6.17) is the heat conduction state; this is a linear static temperature profile between the floor at $T = T_0$ and the ceiling at $T = T_1$,

$$\Theta^{(0)} = 1 - \epsilon y, \quad \vec{v} = 0 \,, \tag{6.24}$$

with the air in rest. Here,

$$\epsilon = \frac{\delta T}{T_0} = \frac{T_0 - T_1}{T_0}$$

is a reduced temperature. Contrary to the former situation, we have now a permanent heat flow through the room, comparable to the case of an open window, and the system will never reach thermal equilibrium as long as $\delta T \neq 0$.

Linearizing the dimensionless equations around (6.24) yields the system

$$\frac{1}{P_r}\partial_t \xi = \nabla^2 \xi + \tilde{R}\,\partial_x \tilde{\Theta}, \tag{6.25a}$$

$$\xi = -\nabla^2 \Psi, \tag{6.25b}$$

$$\partial_t \tilde{\Theta} = \nabla^2 \tilde{\Theta} - \partial_x \Psi \tag{6.25c}$$

with $\tilde{R} = \epsilon R$ and $\tilde{\Theta} = (\Theta - \Theta^{(0)})/\epsilon$. Idealizing the boundary conditions for the vorticity and stream function as[6]

$$\xi = 0, \quad \Psi = 0$$

at all four boundaries, the linear problem (6.25) can be solved exactly with

$$\xi = A \sin(n\pi x)\sin(m\pi y)\,\mathrm{e}^{\lambda t}, \tag{6.26a}$$

$$\Psi = B \sin(n\pi x)\sin(m\pi y)\,\mathrm{e}^{\lambda t}, \tag{6.26b}$$

$$\tilde{\Theta} = C \cos(n\pi x)\sin(m\pi y)\,\mathrm{e}^{\lambda t} \,. \tag{6.26c}$$

Inserting this into (6.25) yields the solvability condition

6 This leads for $\vec{v}$ to the *free boundary conditions*, which actually could be only valid at a plane, free surface.

$$\frac{\lambda^2}{P_r} + \lambda\pi^2(n^2+m^2)(1+1/P_r) + \pi^4(n^2+m^2)^2 - \frac{\tilde{R}\,n^2}{n^2+m^2} = 0 \tag{6.27}$$

for the growth rate λ of the disturbances. For $\tilde{R} > 0$, the polynomial (6.27) possesses two real roots; one of them is always negative. The other has a zero for

$$\tilde{R}^c_{mn} = \frac{\pi^4(n^2+m^2)^3}{n^2} \,. \tag{6.28}$$

Hence the solution (6.24) becomes unstable first with the mode $m = n = 1$ as soon as

$$\tilde{R} \geq \tilde{R}_c = \tilde{R}^c_{11} = 8\pi^4 \approx 779 \,.$$

6.2.5 Frozen time analysis

Now we turn to the fully time dependent diffusion front (6.15) and examine its linear stability. Numerical solutions of the full problem (Figure 6.9) reveal that the front becomes unstable rather soon and the upper boundary $y = L$ can be shifted to infinity. Hence we can treat the problem in the half space

$$0 \leq y \leq \infty$$

and the scaling length L used above is no longer of any relevance. Instead, we take now the space and time scaling

$$(x, y) = (\ell\tilde{x}, \ell\tilde{y}), \quad \tau = \frac{\ell^2}{\kappa}$$

with

$$\ell = \left(\frac{\kappa\nu}{g\alpha\delta T}\right)^{1/3}$$

and obtain a system similar to (6.17) but now only with one parameter, the Prandtl number, left:

$$\frac{1}{P_r}[\partial_t\xi + v_x\partial_x\xi + v_y\partial_y\xi] = \nabla^2\xi + \partial_x\tilde{\Theta}, \tag{6.29a}$$

$$\xi = -\nabla^2\Psi, \tag{6.29b}$$

$$\partial_t\tilde{\Theta} + \vec{v}\cdot\nabla\tilde{\Theta} = \nabla^2\tilde{\Theta} \tag{6.29c}$$

with $\Theta = \epsilon\tilde{\Theta}$. Performing a linear stability analysis about the (scaled) base state (6.15)

$$\tilde{\Theta}^{(0)}(y, t) = 1/\epsilon - \operatorname{erf}(y/2\sqrt{t}) \tag{6.30}$$

results in a linear system with an explicit time dependence. The trick of the "frozen time" analysis, or approximation, is to decouple the time scales of the base state and that of its disturbances. Thus we may consider the time in (6.30) as "frozen" and call it t^*, which brings an additional control parameter into play. Substituting

$$\Psi(x,y,t) = \Phi(y)\,\mathrm{e}^{ikx}\,\mathrm{e}^{\lambda t}, \quad \tilde{\Theta}(x,y,t) = \tilde{\Theta}^{(0)}(y,t^*) + ik\vartheta(y)\,\mathrm{e}^{ikx}\,\mathrm{e}^{\lambda t} \tag{6.31}$$

in (6.29) yields after linearization the generalized eigenvalue problem

$$\frac{\lambda}{P_r}\left(k^2\Phi - \Phi''\right) = -\Phi'''' + 2k^2\Phi'' - k^4\Phi - k^2\vartheta, \tag{6.32a}$$

$$\lambda\vartheta = \vartheta'' - k^2\vartheta + \left(\partial_y\tilde{\Theta}^{(0)}\right)\Phi \tag{6.32b}$$

with

$$\partial_y\tilde{\Theta}^{(0)} = -\frac{1}{\sqrt{\pi t^*}}\,\mathrm{e}^{-\frac{y^2}{4t^*}}$$

and primes denote y-derivatives. The remaining task is to determine the critical time t_c^*, for which the largest eigenvalue λ becomes positive, and the perturbations (6.31) may start to grow exponentially. Since the system (6.32) is a set of linear differential equations with a non-constant coefficient, a numerical solution is in order; see Figure 6.12. Again, the infinite y-domain must be approximated by a finite length L, which, however, must be large compared with the size of the diffusion front (6.30), thus

$$2\sqrt{t^*} \ll L$$

must be checked self-consistently.

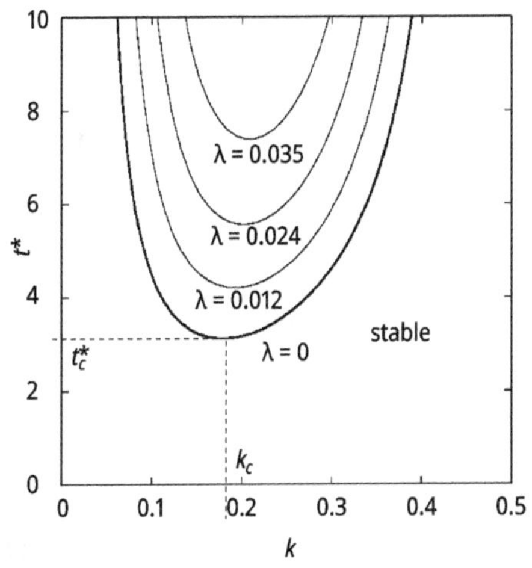

Figure 6.12: Stability chart found numerically by solving (6.32). Modes above the zero-line $\lambda = 0$ (bold) are unstable. Instability occurs first at $t_c^* \approx 3.2$ with $k_c \approx 0.18$.

If we use the realistic values for a room, as stated in Section 6.2.2, and leading to $R \approx 10^{12}$, one finds $\ell \approx 0.6\,\mathrm{mm}$ and $\tau \approx 0.02\,\mathrm{s}$. Thus, the front would be unstable almost immediately at $t_c \approx 0.06\,\mathrm{s}$, and the typical wave length of the growing vorticies is estimated with $\Lambda = 2\pi/k_c \approx 2\,\mathrm{cm}$.

6.3 Convection of a binary mixture

In Section 4.3, we stated the equations (4.39), which describe the state of a binary mixture without computing the velocity field. Now we wish to study the same fluid or gas composition but now in self-organized motion.

6.3.1 Basic equations and boundary conditions

Equations (4.39) have to be completed by the Navier–Stokes equations. For an incompressible mixture and if the Dufour effect is neglected ($D_d = 0$) the basic set reads

$$\partial_t c + \vec{v} \cdot \nabla c = D\,\nabla^2 c + D_s\,\nabla^2 T, \tag{6.33a}$$

$$\partial_t T + \vec{v} \cdot \nabla T = \kappa\,\nabla^2 T, \tag{6.33b}$$

$$\partial_t \vec{v} + (\vec{v} \cdot \nabla)\vec{v} = \nu\,\nabla^2 \vec{v} - \frac{1}{\rho_0}\nabla p - \frac{\rho(T,c)}{\rho_0} g\hat{e}_z, \tag{6.33c}$$

$$\nabla \cdot \vec{v} = 0\,. \tag{6.33d}$$

From here on the vertical coordinate is denoted with z. Again, Boussinesq approximation is assumed and the density is considered as a function of temperature and concentration only in the buoyancy term. A linear state equation according to (4.49),

$$\rho(T,c) = \rho_0(1 - \alpha_T(T - T_0) - \alpha_c(c - c_0)) \tag{6.34}$$

with

$$\alpha_T = -\frac{1}{\rho_0}\left(\frac{\partial \rho}{\partial T}\right)_{c_0,T_0}, \qquad \alpha_c = -\frac{1}{\rho_0}\left(\frac{\partial \rho}{\partial c}\right)_{c_0,T_0}$$

closes the system. (Note here the minus signs in the definition of α_T, α_c).

Let the mixture be confined by two horizontal plates at $z = 0, d$. We apply the boundary conditions from Section 4.3.6 (prescribed temperatures, vanishing diffusion flux) complemented by the rigid wall conditions for the velocity at $z = 0, d$,

$$T(0) = T_0, \quad T(d) = T_1, \tag{6.35a}$$

$$D\,\partial_z c|_{0,d} + D_s \partial_z T|_{0,d} = 0, \tag{6.35b}$$

$$\vec{v}|_{0,d} = 0\,. \tag{6.35c}$$

Here and in the following, we neglect the normally very weak influence of pressure inhomogeneities on the diffusion and heat fluxes.

We use the following scaling to dimensionless variables:

$$(x, z) = d(\tilde{x}, \tilde{z}), \quad t = (d^2/\kappa)\,\tilde{t}, \quad T = \tilde{T}\delta T, \quad c = \tilde{c}\delta T(\alpha_T/\alpha_c), \quad \vec{v} = (\kappa/d)\,\vec{\tilde{v}}\,,$$

with the control parameter

$$\delta T = T_0 - T_1\,.$$

6.3.2 Motionless non-equilibrium state

For $\vec{v} = 0$, we already computed a stationary solution in Section 4.3.6. Temperature and concentration profiles are linear and read

$$\tilde{T}_s(\tilde{z}) = \tilde{T}_0 - \tilde{z}, \quad \tilde{c}_s(\tilde{z}) = \tilde{c}_0 + \beta_c\tilde{z}\,. \tag{6.36}$$

The gradients are related due to (6.35b)

$$\beta_c = s_T\frac{\alpha_c}{\alpha_T}$$

with the Soret coefficient $s_T = D_s/D$, which has been already introduced in Section 4.3.6. The concentration gradient scaled in this way is known in the literature as the (negative) *separation ratio* and often denoted by Ψ:

$$\Psi \equiv -\beta_c = -s_T\frac{\alpha_c}{\alpha_T} = -\frac{D_s}{D}\frac{\alpha_c}{\alpha_T}\,.$$

Note that for a pure fluid and for the cases where the Soret effect plays no role, one has $\Psi = 0$.

6.3.3 Dimensionless equations

The set of basic equations in dimensionless form reads

$$\partial_t\tilde{c} + \vec{\tilde{v}}\cdot\nabla\tilde{c} = L\,\nabla^2\tilde{c} - L\Psi\,\nabla^2\tilde{T}, \tag{6.37a}$$

$$\partial_t\tilde{T} + \vec{\tilde{v}}\cdot\nabla\tilde{T} = \nabla^2\tilde{T}, \tag{6.37b}$$

$$\frac{1}{P_r}(\partial_t\vec{\tilde{v}} + (\vec{\tilde{v}}\cdot\nabla)\vec{\tilde{v}}) = \nabla^2\vec{\tilde{v}} - \nabla\tilde{p} + R(\tilde{T} + \tilde{c})\hat{e}_z, \tag{6.37c}$$

$$\nabla\cdot\vec{\tilde{v}} = 0\,. \tag{6.37d}$$

Four dimensionless groups occur: the two material parameters

$$L = D/\kappa, \quad P_r = \nu/\kappa$$

called *Lewis number* and *Prandtl number*, and the two control parameters

$$R = \frac{g d^3 \alpha_T \delta T}{\nu \kappa}, \quad \Psi = -s_T \frac{\alpha_c}{\alpha_T},$$

the *Rayleigh number* and the *separation ratio*. The Rayleigh number can be simply varied by changing the vertically applied temperature gradient; the separation ratio depends, via the Soret coefficient, on the mean relative concentration; see Figure 4.6 and Figure 6.13.

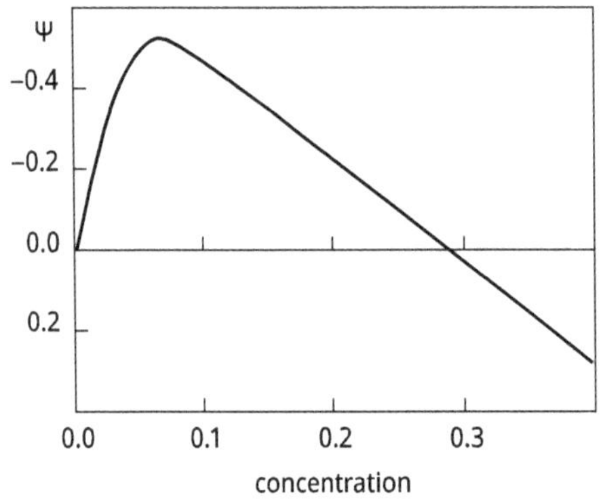

Figure 6.13: Sketch of separation ratio of a water/ethanol mixture as function of the ethanol concentration. For data see [11].

For convenience, we shall leave all tildes in the following. All variables are dimensionless, if not stated otherwise.

6.3.4 Linear stability analysis for large Prandtl number

We treat the problem in 2D, allowing for the description by a stream function,

$$v_x = \partial_z \Phi, \quad v_z = -\partial_x \Phi . \tag{6.38}$$

For the sake of simplicity, we consider only the case $P_r \gg 1$ and neglect the l. h. s. of (6.37c). Applying the curl on (6.37c), one may express Φ through T and c:

$$\nabla^4 \Phi = R\, \partial_x (T + c) . \tag{6.39}$$

Assuming periodic lateral boundary conditions, we may linearize around the steady solution (6.36) according to

$$\begin{aligned} T &= T_s + \vartheta(z)\, \mathrm{e}^{ikx+\lambda t}, \\ c &= c_s + \zeta(z)\, \mathrm{e}^{ikx+\lambda t}, \\ \Phi &= \phi(z)\, \mathrm{e}^{ikx+\lambda t}. \end{aligned} \tag{6.40}$$

Inserting into (6.37a), (6.37b) and (6.39) and keeping only linear terms yields the linear system of ODEs

$$\lambda\zeta + ik\Psi\phi = L(d_z^2 - k^2)\zeta - L\Psi(d_z^2 - k^2)\vartheta, \tag{6.41a}$$

$$\lambda\vartheta + ik\phi = (d_z^2 - k^2)\vartheta, \tag{6.41b}$$

$$(k^4 - 2k^2 d_z^2 + d_z^4)\phi = iRk(\vartheta + \zeta). \tag{6.41c}$$

Together with the boundary conditions at $z = 0, 1$,

$$\vartheta = 0, \quad d_z\zeta - \Psi d_z\vartheta = 0, \quad \phi = d_z\phi = 0\,; \tag{6.42}$$

this system can only be solved by numerical methods. However, for the so-called idealized boundary conditions, namely

$$\vartheta = 0, \quad \zeta = 0, \quad \phi = d_z^2\phi = 0\,, \tag{6.43}$$

an analytical solution is possible, which can at least serve as an estimate for the realistic system. It has the form

$$\vartheta = A_1 \sin\pi z, \quad \zeta = A_2 \sin\pi z, \quad \phi = A_3 \sin\pi z\,.$$

Inserting this into (6.41) yields, after elimination of A_3, a homogeneous system for A_1, A_2, which has the solvability condition

$$\lambda^2 + a\lambda + \omega^2 = 0 \tag{6.44}$$

with

$$\omega^2 = -\frac{Rk^2}{a_k^2}(L + \Psi + L\Psi) + La_k^4, \tag{6.45}$$

$$a = -\frac{Rk^2}{a_k^4}(1 + \Psi) + (L+1)a_k^2 \tag{6.46}$$

and the abbreviation

$$a_k^2 = \pi^2 + k^2\,.$$

As already discussed in Section 5.2.3, there exist two qualitatively different types of instability:

(i) Monotonic instability. A real valued λ changes sign. From (6.44), we have $\omega = 0$, and thus

$$R_M = \frac{(\pi^2 + k^2)^3}{k^2} \frac{L}{L + \Psi + L\Psi} . \tag{6.47}$$

The wave number of the mode becoming unstable for the smallest possible R_M follows from $dR_M/dk = 0$ and turns out to be

$$k_c = \frac{\pi}{\sqrt{2}} , \tag{6.48}$$

independent on L and Ψ. From (6.47), we compute the monotonic instability for

$$R > R_M^c = \frac{27}{4}\pi^4 \frac{L}{L + \Psi + L\Psi} . \tag{6.49}$$

(ii) Oscillatory (Hopf) instability. A complex conjugated pair of λ crosses the real axes if $a = 0$ and $\omega^2 > 0$. The first condition yields

$$R_H = \frac{(\pi^2 + k^2)^3}{k^2} \frac{1 + L}{1 + \Psi} . \tag{6.50}$$

The wave number becoming unstable for the smallest possible R_H is again given by (6.48). From (6.50), we compute the Hopf instability for

$$R > R_H^c = \frac{27}{4}\pi^4 \frac{1 + L}{1 + \Psi} . \tag{6.51}$$

Figure 6.14 shows R_M^c (red line) and R_H^c (blue) both as a function of Ψ for $L = 0.1$. The two curves meet at the codimension-2 (CD2) point, which follows from $R_M^c = R_H^c$ and turns out as

$$\Psi_{CD2} = -\frac{L^2}{1 + L + L^2} .$$

For a liquid mixture L is normally very small (0.01 . . . 0.001) and the codimension-2 point lays almost exactly on the R-axis. Then for practically all negative values of Ψ, an oscillatory instability occurs first while increasing R above its critical value.

The critical (Hopf) frequency along the blue line follows from (6.45) by inserting (6.50) and (6.48) and reads (Figure 6.15)

$$\omega_c = \frac{3}{2}\pi^2 \sqrt{\frac{-L^2 - \Psi(1 + L + L^2)}{1 + \Psi}} . \tag{6.52}$$

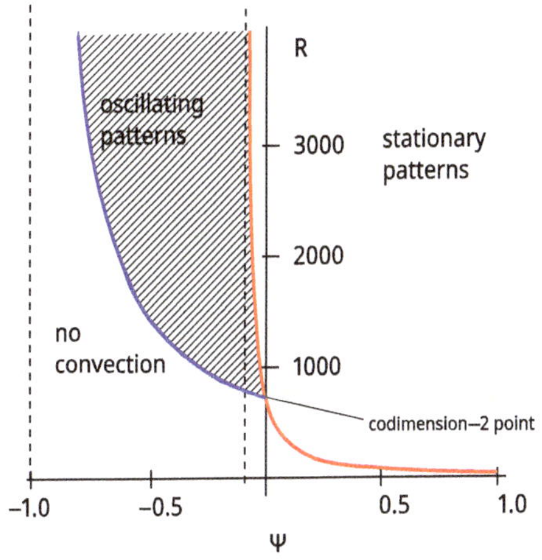

Figure 6.14: Stability diagram of a binary mixture with idealized boundary conditions for $L = 0.1$. In the shaded area, an oscillatory instability occurs. On the right hand side of the red line, monotonic pattern formation is expected. Red and blue lines have the asymptotes $\Psi = -L/(L+1)$ and $\Psi = -1$, respectively. The CD2 point is located very close to the R axis at $\Psi_{CD2} = -L^2/(1+L+L^2)$.

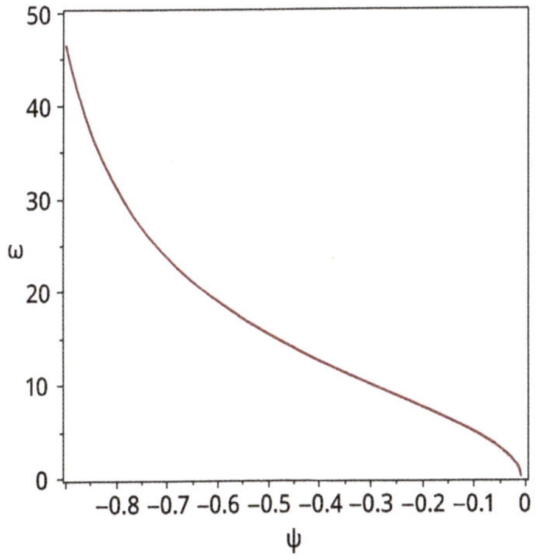

Figure 6.15: Critical frequency (6.52) at the onset of the Hopf instability, $L = 0.1$. The frequency approaches zero at $\Psi = \Psi_{CD2}$ and diverges for $\Psi \to -1$.

6.3.5 The fully nonlinear problem: numerical solutions

To see what kind of patterns evolves and saturates above threshold, we present some numerical solutions of the full set of equations in two spatial dimensions. The method is similar to that already applied in Section 6.2 extended by the transport equation for the concentration field.

To separate the mixed boundary conditions (6.42), it is advantageous to introduce the “combined field”

$$\eta = c - \Psi\, T\,,$$

where $-\nabla\eta$ corresponds to the scaled diffusive mass flux of the solvent. For η the no-flux boundary conditions at $z = 0, 1$ simply read

$$\partial_z \eta = 0 .$$

The boundary conditions for the temperature turn into

$$T(z = 0) = 1, \quad T(z = 1) = 0 .$$

The velocity is again represented by the stream function (now Φ), as introduced in (6.38), and the boundary conditions for Φ and the vorticity ξ are those of Section 6.1.3. The equation for the combined field is found by multiplying (6.37b) with $-\Psi$ and adding (6.37a). The whole set reads

$$\partial_t \eta + v_x \partial_x \eta + v_z \partial_z \eta = L\, \nabla^2 \eta - \Psi\, \nabla^2 T, \tag{6.53a}$$

$$\partial_t T + v_x \partial_x T + v_z \partial_z T = \nabla^2 T, \tag{6.53b}$$

$$\frac{1}{P_r}(\partial_t \xi + v_x \partial_x \xi + v_z \partial_z \xi) = \nabla^2 \xi + R(\partial_x \eta + (1 - \Psi)\partial_x T), \tag{6.53c}$$

$$\nabla^2 \Phi = -\xi . \tag{6.53d}$$

In the following, we show solutions for $\Psi = -1/2$, $L = 0.01$, $P_r = 7$ in the oscillatory regime. The code uses a 400×40 FD-grid, corresponding to an aspect ratio of $\Gamma = 10$. Figure 6.16 shows the formation of a traveling wave pattern. After an initial phase, the pattern travels with a constant speed to the left-hand side without further changes of shape.

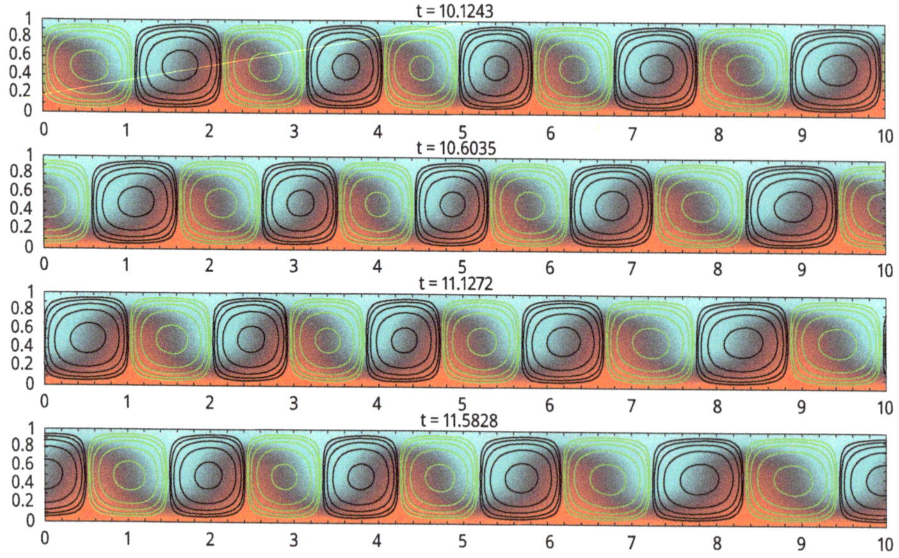

Figure 6.16: Numerical computations of the evolution of a binary mixture. Contour lines show stream lines Φ = const.; colors correspond to the temperature field (bathroom convention). Parameters: $R = 3000$, $\Psi = -0.5$, $L = 0.01$, $P_r = 7$.

The earlier states of pattern evolution depend here strongly on the initial conditions. To demonstrate this, we plot the temperature of the midpoint

$$T_M(t) = T(x = \Gamma/2, z = 1/2, t)$$

over time in Figure 6.17. If we start with a small $R = 2000$ and switch then to 3000 after $t = 9$, a traveling wave state, as that shown in Figure 6.16, emerges rather fast (blue line). If we switch a bit sooner ($t = 8$), the same TW-state but with a smaller speed is found (yellow). If $R = 3000$ directly at $t = 0$, an almost stationary wave state is found (green), which accelerates in the following very slowly, probably to the same speed, than that of the other states. This behavior is even more pronounced if we take at $t = 0$ a larger $R = 5000$ (purple), Figure 6.18.

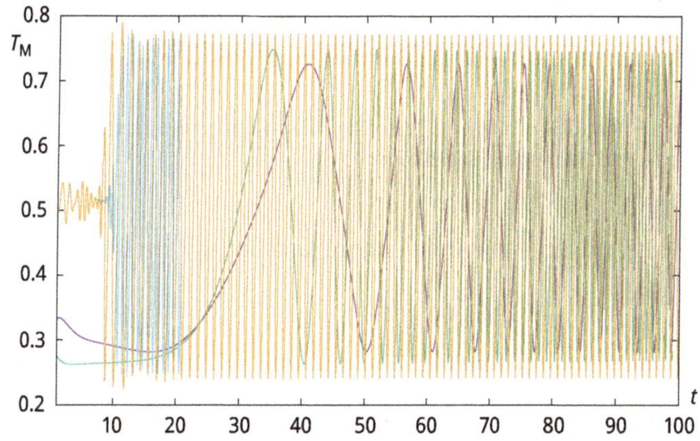

Figure 6.17: Temperature at the midpoint of the layer over time for different initial conditions; see text.

6.4 Problems

6.4.1 Internal waves in fluids

If we neglect viscosity and thermal diffusivity, a fluid is described by the equations

$$\rho \frac{d\vec{v}}{dt} = -\nabla P + \vec{f}, \tag{1a}$$

$$\frac{d\rho}{dt} = -\rho\, \nabla \cdot \vec{v}, \tag{1b}$$

$$C\rho \frac{dT}{dt} = -P\, \nabla \cdot \vec{v} \tag{1c}$$

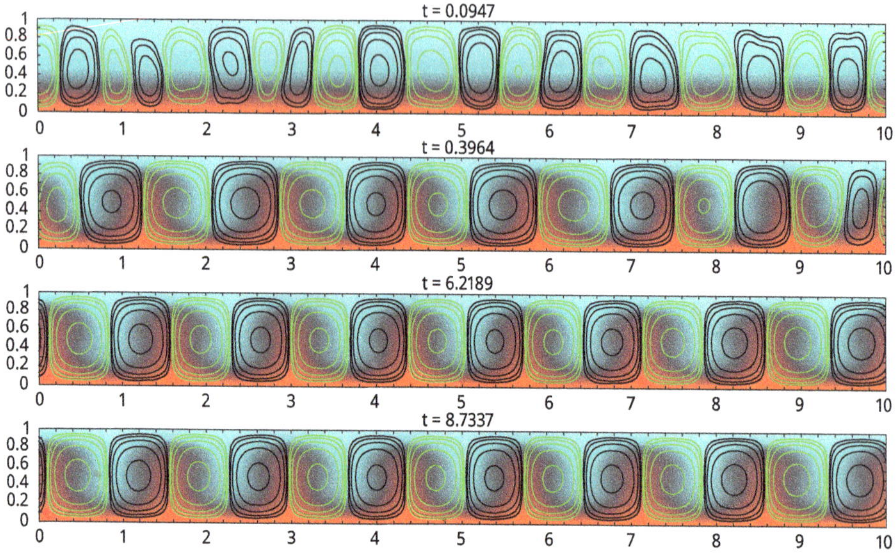

Figure 6.18: Same as Figure 6.16 but now for $R = 5000$ directly from $t = 0$. An almost steady pattern occurs rapidly after $t = 6$, which then accelerates its traveling speed continuously.

with the (constant) specific heat C. Pressure, temperature, and density are linked by a thermal state equation

$$P = h(\rho, T)\,. \tag{2}$$

For the external force $\vec{f}$, you may assume $\nabla \times \vec{f} = 0$ (potential force).

a) Linearization. Assume a given stationary state with

$$\vec{v}_S = 0, \quad \nabla P_S(\vec{r}) = \vec{f}(\vec{r}), \quad \rho = \rho_S(\vec{r}), \quad T = T_S(\vec{r})\,.$$

Insert the perturbation ansatz

$$\rho = \rho_S + \tilde{\rho}, \quad T = T_S + \tilde{T}, \quad P = P_S + \tilde{P}$$

into (1). Show that you obtain the system

$$\rho_S \dot{\vec{v}} = -\nabla \tilde{P}, \tag{3a}$$

$$\dot{\tilde{\rho}} = -\rho_S \nabla \cdot \vec{v} - \vec{v} \cdot \nabla \rho_S, \tag{3b}$$

$$\dot{\tilde{T}} = -\frac{P_S}{C\rho_S} \nabla \cdot \vec{v} - \vec{v} \cdot \nabla T_S \tag{3c}$$

if you neglect all nonlinear terms in the perturbations.

b) Sound waves. Assume a space independent, homogeneous stationary state $\rho_s = \rho_0 =$ const., $T_s = T_0 =$ const., $P_s = P_0 =$ const., and neglect $\vec{f}$. Take for (2) the perfect gas equation

$$P = R^* T\rho \tag{4}$$

with $R^* = R/M \approx 286\,\text{J/(kg·K)}$ for air, the gas constant R, and the mole mass M. With (4) you can write

$$\tilde{P} = R^*(T_0\tilde{\rho} + \rho_0\tilde{T})\,. \tag{5}$$

Why? Use the linear system (3) and derive a wave equation for $\tilde{P}$ of the form

$$\Delta\tilde{P} - \frac{1}{u_p^2}\frac{\partial^2\tilde{P}}{\partial t^2} = 0\,. \tag{6}$$

What do you find for the phase velocity u_p of sound waves? Compute u_p with $C = 717\,\text{J/kg·K}$ and $T_0 = 300\,\text{K}$. What do you obtain if the air is assumed to be isothermal (Newton's solution)? Which value is the better one?

c) Internal gravity waves. Sound waves are longitudinal waves and can only propagate in a compressible medium. On the other hand, transversal waves may also exist in incompressible fluids. Assume now

$$\nabla\cdot\vec{v} = 0 \tag{7}$$

and an isothermal fluid with $T = T_0$, being in the constant gravity field $\vec{f} = -\rho g\hat{e}_z$ with $g = 9.81\,\text{m/s}^2$. In the Boussinesq approximation, the density is assumed being constant, except for the external force term, where it accounts for buoyancy. Then instead of (1) we obtain the system

$$\frac{d\vec{v}}{dt} = -\frac{1}{\rho_0}\nabla P - \frac{\rho}{\rho_0}g\hat{e}_z, \tag{8a}$$

$$\frac{d\rho}{dt} = 0\,. \tag{8b}$$

Find the general hydrostatic solution ($\vec{v} = 0$) ρ_s, P_s of (8), and show that ρ_s can be an arbitrary function of z. Show also that the linearized system with respect to ρ_s and P_s has the form

$$\dot{\vec{v}} = -\frac{1}{\rho_0}\nabla\tilde{P} - \frac{\tilde{\rho}}{\rho_0}g\hat{e}_z, \tag{9a}$$

$$\dot{\tilde{\rho}} = -w\partial_z\rho_s\,, \tag{9b}$$

where $\vec{v} = (u, v, w)$. Finally, derive a wave equation from (9) for internal gravity waves that has the form

$$\Delta\ddot{w} + N^2\Delta_2 w = 0\,, \tag{10}$$

where $\Delta_2 = \partial_{xx} + \partial_{yy}$ is the horizontal Laplacian.

Hint: Eliminate the pressure by forming twice the curl on equation (9a) and use the identity

$$\nabla \times (\nabla \times \vec{A}(\vec{r})) = \nabla(\nabla \cdot \vec{A}(\vec{r})) - \Delta\vec{A}(\vec{r}),$$

valid for an arbitrary vector field $\vec{A}$.

What do you obtain for the so-called *Brunt–Väisälä frequency*[7] N (see Figure 6.19)? Compute the dispersion relation for plane waves

$$w(\vec{r}, t) = A\, e^{i(\vec{k}\vec{r}-\omega t)} \tag{11}$$

from (10). Can N^2 become negative? And what happens if so?

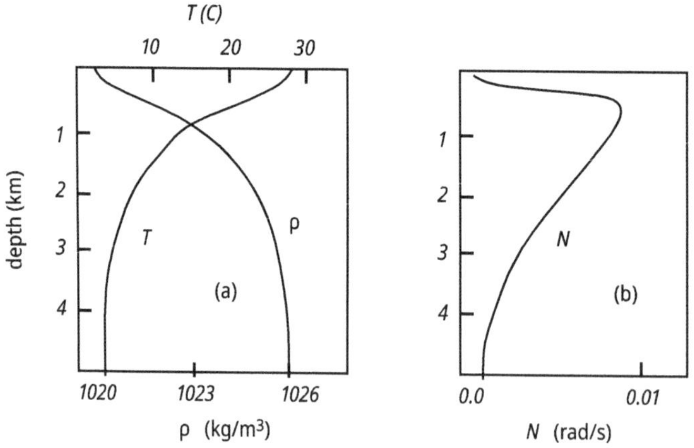

Figure 6.19: Sketch of (a) temperature and density, (b) Brunt–Väisälä frequency in the ocean. For data see [12].

6.4.2 Convection in a porous medium

Many substances, such as rocks, soil, biological tissues (e. g., bones, wood, cork), cements or ceramics can be considered as porous media. A porous medium is a solid material

7 V. Väisälä, 1899–1969, Finnish meteorologist, D. Brunt, 1886–1965, British meteorologist.

containing connected pores. The pores can be filled with water and a flow through a porous medium is described by Darcy's law[8]

$$\vec{q} = -\frac{K_0}{\eta}(\nabla P + g\rho\,\hat{e}_z)\,. \tag{12}$$

Here, $\vec{q}(\vec{r},t)$ is the volumetric flux [m/s] of the fluid, η its dynamic viscosity [Pa s], and ρ its density [kg/m^3]. The pressure is denoted by $P(\vec{r},t)$ and g accounts for gravity acceleration. A porous medium is characterized by its permeability K_0 [m^2] and its porosity n. Note that $\vec{q}$ is related to the fluid velocity through

$$\vec{q} = n\vec{v}\,.$$

The porosity n is defined as the ratio of the volume of empty spaces (the pores) to the total volume of the medium, therefore $0 \le n \le 1$.

For a non-isothermal medium, in addition the temperature equation must be considered

$$\frac{\partial T}{\partial t} + \frac{1}{n}\vec{q}\cdot\nabla T = \kappa\nabla^2 T \tag{13}$$

with the thermal diffusivity κ. Then the density ρ in (12) normally depends on temperature according to

$$\rho = \rho_0[1 - \alpha(T - T_0)]\,, \tag{14}$$

where $\alpha = -(\partial\rho/\partial T)/\rho_0$ is the thermal expansion coefficient.

a) The system. Consider a 2D porous medium with height d and an infinite lateral extension (Figure 6.20). The temperatures on the lower and upper borders are T_0, T_1, respectively, with $T_0 > T_1$. Compute the stationary temperature distribution $T^{(0)}(z)$ and the pressure $P^{(0)}(z)$ of the motionless state $\vec{q} = 0$ by solving (12), (13) with (14).

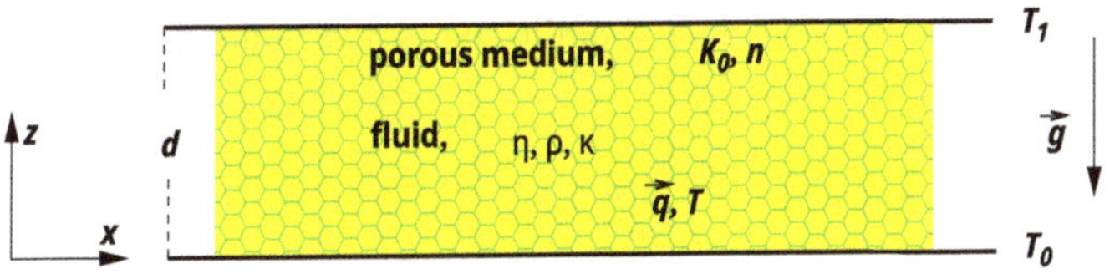

Figure 6.20: Porous medium heated from below.

8 Henry Darcy, French engineer, 1803–1858, working mainly in hydraulics.

b) Nondimensionalization. Write the coupled basic equation (12), (13) in nondimensional form. Use the following scaling (all variables marked with a tilde are dimensionless):

$$(x,z) = (\tilde{x},\tilde{z}) \cdot d, \quad t = \tilde{t} \cdot \tau, \quad \vec{q} = \frac{d}{\tau} \cdot n \cdot \vec{\tilde{q}}, \quad T = \delta T \cdot \tilde{T}$$

with $\delta T = T_0 - T_1$ and $\tau = d^2/\kappa$. By the help of relation (14), you should obtain the dimensionless set

$$\vec{\tilde{q}} = -\tilde{\nabla}\tilde{P} + R_a\,(\tilde{T} - \tilde{T}_0)\,\hat{e}_z, \tag{15a}$$

$$\frac{\partial \tilde{T}}{\partial \tilde{t}} + \vec{\tilde{q}} \cdot \tilde{\nabla}\tilde{T} = \tilde{\nabla}^2 \tilde{T}\,. \tag{15b}$$

What do you find for $\tilde{P}$ and what for the Rayleigh number R_a?

c) Stream function. Show that

$$\vec{\tilde{q}} = \tilde{\nabla} \times (\tilde{\Phi}(\tilde{x},\tilde{z},\tilde{t})\,\hat{e}_y) \tag{16}$$

fulfills the incompressibility condition

$$\nabla \cdot \vec{v} = 0\,.$$

Derive an equation for the stream function $\tilde{\Phi}$ introduced in (16) by applying the curl on (15a).

d) Linear stability. Continue with the set

$$\tilde{\nabla}^2\tilde{\Phi} = R_a \partial_{\tilde{x}}\tilde{T}, \tag{17a}$$

$$\frac{\partial \tilde{T}}{\partial \tilde{t}} + \partial_{\tilde{x}}\tilde{\Phi}\partial_{\tilde{z}}\tilde{T} - \partial_{\tilde{z}}\tilde{\Phi}\partial_{\tilde{x}}\tilde{T} = \tilde{\nabla}^2\tilde{T} \tag{17b}$$

and linearize it with respect to the perturbations $\tilde{\Theta}$ of the conductive state $\tilde{T}^{(0)}(\tilde{z})$ computed in **a)**,

$$\tilde{T}(\tilde{x},\tilde{z},\tilde{t}) = \tilde{T}_0 - \tilde{z} + \tilde{\Theta}(\tilde{x},\tilde{z},\tilde{t})\,. \tag{18}$$

Show that the linear system is solved by

$$\tilde{\Theta} = A\sin(\ell\pi\tilde{z})\,e^{ik\tilde{x}}\,e^{\lambda\tilde{t}}, \quad \tilde{\Phi} = B\sin(\ell\pi\tilde{z})\,e^{ik\tilde{x}}\,e^{\lambda\tilde{t}} \tag{19}$$

and compute the growth rate $\lambda(k,\ell)$. Plot λ as a function of k for $\ell = 1, 2$ and $R_a = 50$.

What do you find for the smallest (critical) value R_a, where the conductive state becomes unstable and convection sets in ($\lambda = 0$)? Determine k and ℓ for this case.

Compute the minimal temperature difference δT (in Kelvin) needed for the onset of convection in a porous medium with

$$K_0 = 10^{-11}\,\mathrm{m}^2, \quad n = 0.2, \quad d = 100\,\mathrm{m}\,.$$

Take (i) water, (ii) air for the fluid. For values of ν, κ, α, see Appendix A.

Part II: **Kinetic gas theory**

Ludwig Eduard Boltzmann (1844–1906). Founder of the statistical definition of entropy.
Picture from 1902, unknown author, source: Wikimedia.

https://doi.org/10.1515/9783110794137-008

7 Many particles and probability distribution

The phenomenological theory presented so far has some similarity to a "black-box approach": a thermodynamic system is considered as a macroscopic entity, described by as few as possible independent state variables, which may become functions of time and space if the system is out of global equilibrium. But everybody interested in sciences naturally wants to know (and already has at least an idea about) what is going on inside the black box: atoms or molecules (traditionally called particles) may interact by forces or collisions and macroscopically form the known states of matter, such as solid, liquid or gaseous. Eventually, after a certain kind of averaging or coarse graining, the particles, their mutual interactions, and their response to external forces must be responsible for all objects forming our macroscopic world and their complete macroscopic behavior observed in experiments and described by phenomenological thermodynamics.

In the remaining chapters, we wish to have a brief look inside the box. We shall mainly consider the most simple case: a perfect one-component gas constituted of a constant number of single atoms (particles) that follow the rules of classical mechanics. The particles can be exposed to external forces (e. g., gravity). The only allowed interactions between them are binary collisions and collisions with the walls limiting their free space. In both cases, mass, momentum, and energy must be conserved. The conservation of energy implies a Hamiltonian dynamics without any friction on the particle level. As a consequence, all microscopic processes are reversible in time.

7.1 Phase space

Consider N particles, all with equal mass m, at locations $\vec{r}_i(t)$, and with velocities $\vec{v}_i(t)$. The six-dimensional phase space (sometimes called μ-space) is spanned by the three space axes ($\vec{r}$) and the three velocity axes ($\vec{v}$).[1] Each particle has a defined position in phase space that moves with time t along its trajectory.

To find classical particle motion under given internal and external forces, one must resort to Newton's law of motion, nowadays often written in the form of a set of $6N$ first-order ordinary differential equations:

$$\frac{d\vec{r}_i}{dt} = \frac{1}{m}\frac{\partial H}{\partial \vec{v}_i}, \tag{7.1a}$$

$$\frac{d\vec{v}_i}{dt} = -\frac{1}{m}\frac{\partial H}{\partial \vec{r}_i} \tag{7.1b}$$

1 In statistical mechanics, normally the three momentum components $\vec{p}_i$ are considered, but since in our case $\vec{p}_i = m\vec{v}_i$ with the constant particle mass m, we can also use $\vec{v}_i$ as dependent variables.

https://doi.org/10.1515/9783110794137-009

with the Hamilton function

$$H(\vec{r}_i, \vec{v}_i) = \frac{m}{2}\sum_i^N v_i^2 + \sum_{i,j<i}^N U(r_{ij}) + U_e(\vec{r}_i)$$

and $r_{ij} \equiv |\vec{r}_i - \vec{r}_j|$. Particle interaction, also collisions, are included in the pair potential $U(r_{ij})$; $U_e(\vec{r})$ is the potential of the external forces $\vec{F}_e = -\nabla U_e$.

For simple, normally not very interesting or even trivial cases, the trajectories can be given analytically. For example, for non-interacting ($U = 0$), vertically free moving particles (one dimension) in a constant gravitation field ($U_e = mgz$), one finds $H(z, v) = \frac{m}{2}v^2 + mgz$ and the solution of (7.1) reads

$$z(t) = z_0 - \frac{1}{2}gt^2 + v_0 t, \quad v(t) = v_0 - gt\,.$$

Eliminating t yields for the trajectories

$$z(v) = z_0 - \frac{1}{2g}(v^2 - v_0^2)$$

with the initial values z_0, v_0 for position and velocity (Figure 7.1).

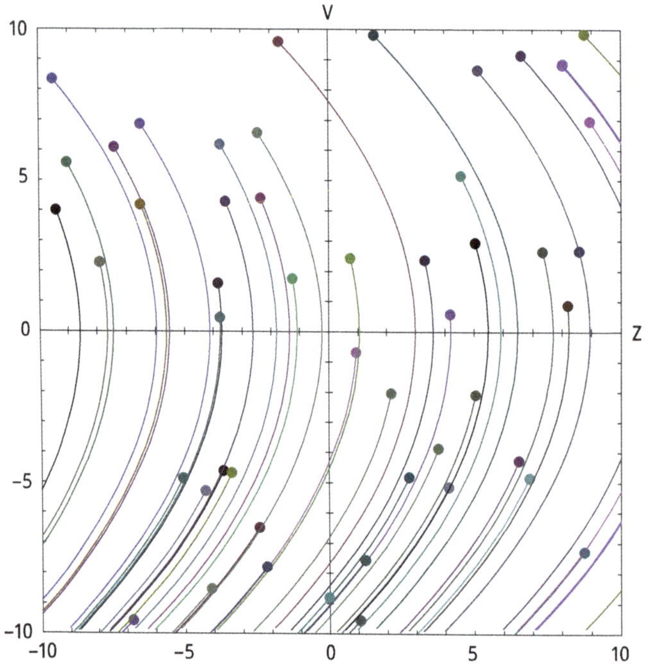

Figure 7.1: Section of the 2D-phase space for a 1D-vertically moving particle under a constant gravitation force. The trajectories begin at the dots with randomly chosen initial conditions and are passed downwards if time goes on.

However, for the problems we are concerned with, N is huge for one mole of a gas around $6 \cdot 10^{23}$, and it is hopeless even to think about a solution of (7.1). On the other hand, it would be also useless to know position and velocity of some $6 \cdot 10^{23}$ particles at each time. Of particular interest in this situation is to compute the mean and expectation values, leading directly into the field of statistics.

7.2 Distribution function

7.2.1 Probability density

A central item of kinetic gas theory is the *distribution function* or *probability density function* (PDF)

$$f(\vec{r}, \vec{v}, t), \tag{7.2}$$

which is defined as the probability to find a number of dN particles at position $\vec{r}$ with velocity $\vec{v}$ in the (small) interval $d^3\vec{r}\; d^3\vec{v} = dx\, dy\, dz\, dv_x dv_y dv_z$:

$$dN = f(\vec{r}, \vec{v}, t)\; d^3\vec{r}\; d^3\vec{v}\,. \tag{7.3}$$

Note that the dimension of f is one over volume and velocity cubed. From f and its moments all thermodynamic variables of interest can be computed. We use in the following the abbreviation for the mean value

$$\langle h \rangle \equiv \int_{-\infty}^{\infty} d^3\vec{v}\; h f(\vec{r}, \vec{v}, t)\,. \tag{7.4}$$

Here, h can be a function of $\vec{r}$, $\vec{v}$, and t. Since $\vec{v}$ is integrated out, $\langle h \rangle$ depends only on $\vec{r}$ and on t. If the gas is in global thermal equilibrium, $f = f(\vec{v})$ only, and $\langle h \rangle$ is a constant with respect to space and time. The dimension of $\langle h \rangle$ is equal to that of h divided by volume, i. e. the density of h.

Let us list the most important moments of f:

(1) **Particle density:**

$$n(\vec{r}, t) = \langle 1 \rangle = \int_{-\infty}^{\infty} d^3\vec{v}\, f\,. \tag{7.5}$$

This is the number of particles in $d^3\vec{r}$ with arbitrary velocities. From there the density

$$\rho(\vec{r}, t) = m\, n(\vec{r}, t) = \langle m \rangle$$

follows with m as the mass of one particle.

(2) **Mean velocity:**

$$\vec{u}(\vec{r},t) = \frac{1}{n}\langle\vec{v}\rangle = \frac{1}{n}\int_{-\infty}^{\infty} d^3\vec{v}\,\vec{v}\,f\,. \tag{7.6}$$

We split

$$\vec{v} = \vec{u}(\vec{r},t) + \vec{w}(\vec{r},t) \tag{7.7}$$

and call the deviations $\vec{w}$ from the mean the *random velocities*. From (7.6), it is clear that

$$\langle\vec{w}\rangle = 0\,.$$

(3) **Kinetic energy:** With (7.7) and (7.6), we can write for the kinetic energy density of the gas:

$$e = \frac{1}{2}m\langle v^2\rangle = \frac{1}{2}m(nu^2 + \langle w^2\rangle + 2\langle\vec{u}\cdot\vec{w}\rangle)\,. \tag{7.8}$$

Due to

$$\langle\vec{u}\cdot\vec{w}\rangle = \vec{u}\cdot\langle\vec{w}\rangle = 0,$$

we have found an expression analogue to (2.27)

$$e = e_k + u_I$$

with the kinetic energy density of the mean flow

$$e_k = \frac{1}{2}n\,mu^2 = \frac{1}{2}\rho\,u^2$$

and the internal energy density

$$u_I = \frac{1}{2}m\langle w^2\rangle\,. \tag{7.9}$$

7.2.2 Pressure and equipartition theorem

Consider a perfect gas in global equilibrium ($\vec{u} = 0$) in a cube with side length L. The pressure generated by a perfect gas on a sidewall, say at $x = L$ with area $A = L^2$ originates from the elastic collisions of the particles and the wall. The force generated by a single particle collision reads

$$F = -m\frac{\Delta w}{\delta t},$$

where $\Delta w = (\vec{w}_2 - \vec{w}_1)_x = -2w_{1x}$ and $\vec{w}_1$ as velocity before and $\vec{w}_2$ after the collision; see Figure 7.2. For the pressure caused by a single particle we thus have

$$p_s = \frac{F}{A} = \frac{2mw_{1x}}{A\Delta t} .$$

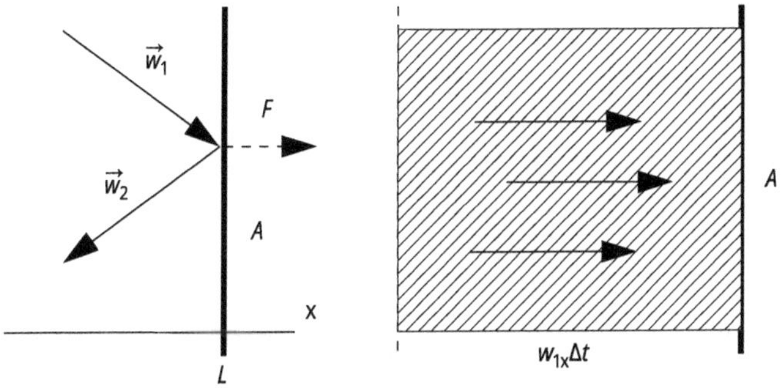

Figure 7.2: Left: a particle is reflected by an elastic collision at the wall, $w_{2x} = -w_{1x}$, $w_{2y} = w_{1y}$, $w_{2z} = w_{1z}$. Right: A particle inside the volume $\Delta V = Aw_{1x}\Delta t$ reaches the wall within the time interval Δt.

To find the pressure caused by all particles, we must consider those particles that can reach the wall within time Δt. They are in a cuboid of volume $\Delta V = w_{1x}A\Delta t$. Averaging over all particles we find

$$p = \int_0^\infty dv_x \int_{-\infty}^\infty dv_y \int_{-\infty}^\infty dv_z \frac{2mw_{1x}}{A\Delta t}\Delta V\, f = m \int_{-\infty}^\infty dv_x \int_{-\infty}^\infty dv_y \int_{-\infty}^\infty dv_z w_{1x}^2 f$$
$$= m\langle w_x^2\rangle .$$

If the gas is isotropic, one must have

$$\langle w_x^2\rangle = \langle w_y^2\rangle = \langle w_z^2\rangle = \frac{1}{3}\langle w^2\rangle ,$$

and the pressure can be written in the form

$$p = \frac{1}{3}m\langle w^2\rangle . \tag{7.10}$$

From the perfect gas equation, $\langle w^2\rangle$ is also directly related to the temperature according to

$$k_B T = \frac{m}{\rho}p = \frac{1}{3}\frac{m}{n}\langle w^2\rangle , \tag{7.11}$$

where $k_B = R/A_v$ is Boltzmann's constant, R the universal gas constant, and A_v Avogadro's number. Comparing with (7.9), we derive the important *equipartition theorem*

$$u_I = \frac{f_d}{2} n k_B T, \tag{7.12}$$

where f_d is the number of degrees of freedom for a single particle. For our monatomic gas, we have $f_d = 3$, and equating (7.9), (7.12) yields

$$n k_B T = \frac{1}{3} m\langle w^2\rangle = m\langle w_x^2\rangle = m\langle w_y^2\rangle = m\langle w_z^2\rangle . \tag{7.13}$$

For a perfect gas, random particle motion is linearly linked to temperature, pressure, and internal energy. All three quantities have, up to prefactors, the same meaning.

7.2.3 Maxwell–Boltzmann distribution

First let us assume a perfect gas in global equilibrium and $f = f(\vec{v})$ only. We shall determine f for this case. Since we are interested later in a distribution allowing for a mean flow $\vec{u} \neq 0$, we present the derivation using random velocities $\vec{w}$ introduced in (7.7).

Since the components of $\vec{w}$ are independent, f should factorize

$$f(\vec{w}) = f_x(w_x) f_y(w_y) f_z(w_z) .$$

We can look for the variation of f if $\vec{w}$ varies:

$$\delta f = \frac{\partial f}{\partial w_x}\delta w_x + \frac{\partial f}{\partial w_y}\delta w_y + \frac{\partial f}{\partial w_z}\delta w_z = f\left(\frac{1}{f_x}\frac{\partial f_x}{\partial w_x}\delta w_x + \frac{1}{f_y}\frac{\partial f_y}{\partial w_y}\delta w_y + \frac{1}{f_z}\frac{\partial f_z}{\partial w_z}\delta w_z\right). \tag{7.14}$$

But if the gas is isotropic, f should only depend on powers of w^2 (rotationally invariant in $\vec{w}$-space), and if the variation of $\vec{w}$ is in such a way, that w^2 is kept constant, i. e.,

$$w_x \delta w_x + w_y \delta w_y + w_z \delta w_z = 0 , \tag{7.15}$$

then δf must vanish. From (7.14), this means

$$\frac{1}{f_x}\frac{\partial f_x}{\partial w_x}\delta w_x + \frac{1}{f_y}\frac{\partial f_y}{\partial w_y}\delta w_y + \frac{1}{f_z}\frac{\partial f_z}{\partial w_z}\delta w_z = 0 . \tag{7.16}$$

Thus we can consider (7.16) as a variational problem with the constraint (7.15) (or vice versa). Applying the method of Lagrange multipliers, we multiply (7.15) with the Lagrange parameter $m\beta$ and add it to (7.16):

$$\frac{1}{f_x}\frac{\partial f_x}{\partial w_x}\delta w_x + \frac{1}{f_y}\frac{\partial f_y}{\partial w_y}\delta w_y + \frac{1}{f_z}\frac{\partial f_z}{\partial w_z}\delta w_z + m\beta(w_x \delta w_x + w_y \delta w_y + w_z \delta w_z) = 0 ,$$

or

$$\left(\frac{1}{f_x}\frac{\partial f_x}{\partial w_x} + m\beta w_x\right)\delta w_x + \left(\frac{1}{f_y}\frac{\partial f_y}{\partial w_y}\delta w_y + m\beta w_y\right)\delta w_y$$
$$+ \left(\frac{1}{f_z}\frac{\partial f_z}{\partial w_z}\delta w_z + m\beta w_z\right)\delta w_z = 0\,. \qquad (7.17)$$

Now the variations δw_i are independent of each other, and to fulfill (7.17) each bracket must vanish independently:

$$\frac{\partial f_i}{\partial w_i} + m\beta w_i f_i = 0\,,$$

where $i = x, y, z$. The solution of each equation reads (integration by separation)

$$f_i(w_i) = A_i e^{-\frac{m\beta}{2}w_i^2}\,,$$

and finally with $\vec{w} = \vec{v} - \vec{u}$,

$$f(\vec{v}) = A e^{-\frac{m\beta}{2}(\vec{v}-\vec{u})^2}\,, \quad A = A_x A_y A_z\,. \qquad (7.18)$$

As last steps, we determine A from the normalization (7.5) with $n = N/V$ (global equilibrium)

$$A = \frac{N}{V}\left(\frac{m\beta}{2\pi}\right)^{3/2}\,,$$

and β from (7.13)

$$\beta = \frac{1}{k_B T}$$

and receive

$$\boxed{f(\vec{v}) = \frac{N}{V}\left(\frac{m}{2\pi k_B T}\right)^{3/2} \exp\left[-\frac{m(\vec{v}-\vec{u})^2}{2k_B T}\right]}\,. \qquad (7.19)$$

This is the *Maxwell–Boltzmann distribution*, also called *Maxwellian*, for a monatomic perfect gas with N particles in a volume V in global equilibrium; see Figure 7.3.

7.2.4 Local equilibrium distribution

It is straightforward to do the same steps leading from Chapter 1 to Chapter 2: assuming local equilibrium, instead of global one, and take the temperature and the mean velocity $\vec{u}$ as slowly varying functions in space and time. The homogeneous particle density

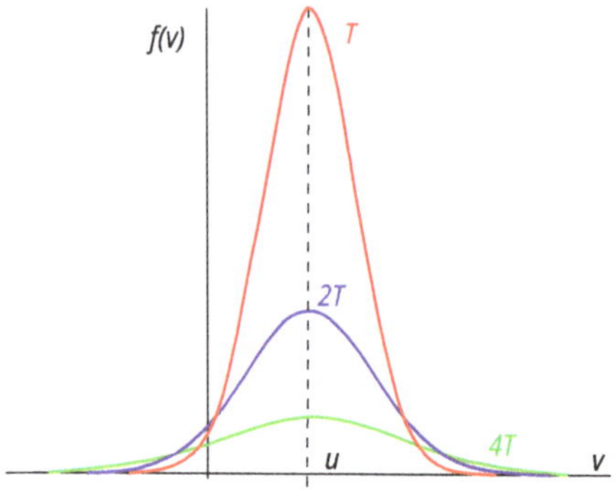

Figure 7.3: Sketch of the Maxwell–Boltzmann distribution (7.19) for different temperatures; section in direction of $\vec{u}$, $\vec{v} = v\hat{u}$.

N/V must be replaced by $n(\vec{r},t)$ or $\rho(\vec{r},t)/m$. From (7.19) this results into the now space, velocity, and time dependent PDF

$$f(\vec{r},\vec{v},t) = n(\vec{r},t)\left(\frac{m}{2\pi k_B T(\vec{r},t)}\right)^{3/2} \exp\left[-\frac{m(\vec{v}-\vec{u}(\vec{r},t))^2}{2k_B T(\vec{r},t)}\right], \tag{7.20}$$

which is called *local Maxwell–Boltzmann distribution.*

7.2.5 Multi-component systems

Although the focus of the statistic methods presented here is on a one-component gas, we shall give a brief outline on the treatment of a gas mixture. If the gases are monatomic, the only distinguishing feature is the particle mass, which we assume to be different. The extension is straightforward, instead of one PDF, one needs now a distribution function for each component,

$$f^k(\vec{r},\vec{v},t),$$

which describes the number density of particles of component k in $d^3\vec{r}\,d^3\vec{v}$. We adopt the notation of Section 7.2.1 and define the moment of a function h with

$$\langle h\rangle^k \equiv \int_{-\infty}^{\infty} d^3\vec{v}\; h f^k(\vec{r},\vec{v},t)\,. \tag{7.21}$$

Then the partial particle density of component k is given as

$$n_k(\vec{r},t) = \langle 1\rangle^k = \int_{-\infty}^{\infty} d^3\vec{v}\, f^k$$

and

$$n = \sum_k^K n_k$$

as total particle density. The partial mass density is

$$\rho_k(\vec{r}, t) = m_k\, n_k(\vec{r}, t) = \langle m_k \rangle^k$$

with m_k as the mass of one particle of component k. The total density reads

$$\rho = \sum_k^K \rho_k \,.$$

The velocity of component k is found from

$$\vec{u}_k(\vec{r}, t) = \frac{1}{n_k} \langle \vec{v} \rangle^k = \frac{1}{n_k} \int_{-\infty}^{\infty} d^3\vec{v}\, \vec{v}\, f^k \tag{7.22}$$

and the barycentric velocity from

$$\vec{u} = \frac{1}{\rho} \sum_k^K \rho_k \vec{u}_k \,. \tag{7.23}$$

The kinetic energy is written as

$$e = \frac{1}{2} \sum_k^K m_k \langle v^2 \rangle^k = \frac{1}{2} \sum_k^K m_k (n_k u^2 + 2\vec{u} \cdot \langle \vec{w} \rangle^k + \langle w^2 \rangle^k) \,, \tag{7.24}$$

where we substituted $\vec{v} = \vec{u} + \vec{w}$. Note that the definition of the random velocity $\vec{w}$ differs from the single component gas. It now denotes the deviations from the barocentric velocity and not from the mean velocity $\vec{u}_k$. For the second expression in the brackets on the r. h. s., we have

$$\vec{u} \cdot \sum_k^K m_k \langle \vec{v} - \vec{u} \rangle = \vec{u} \cdot \sum_k^K (n_k m_k \vec{u}_k - n_k m_k \vec{u}) = \vec{u} \cdot \left(\sum_k^K \rho_k \vec{u}_k - \rho \vec{u} \right) = 0 \,,$$

where we used (7.23). Again, the kinetic energy splits into two parts,

$$e = e_k + u_I$$

with the kinetic energy density

$$e_k = \frac{1}{2} \sum_k^K n_k\, m_k u^2 = \frac{1}{2} \rho\, u^2$$

and the internal energy density

$$u_I = \frac{1}{2}\sum_k^K m_k \langle w^2 \rangle^k . \tag{7.25}$$

If the mixture is in (local) equilibrium, a Maxwell–Boltzmann distribution applies for each component according to

$$f^k(\vec{r}, \vec{v}, t) = n_k(\vec{r}, t)\left(\frac{m_k}{2\pi k_B T(\vec{r}, t)}\right)^{3/2} \exp\left[-\frac{m_k(\vec{v} - \vec{u}_k(\vec{r}, t))^2}{2k_B T(\vec{r}, t)}\right] . \tag{7.26}$$

Inserting this into (7.25), we finally find

$$u_I = \frac{3}{2} n k_B T ,$$

the same result as for a one-component gas.

7.3 Entropy

Looking into the black box, the first connection of statistics with thermodynamics we recognized was the equivalence of microscopic particle random velocities and macroscopic temperature (or pressure) according to the equipartition theorem (7.13). The next link is a rather natural description of entropy in terms of statistical variables and disorder.

7.3.1 Microstates, macrostates, and multiplicity

They key idea of Boltzmann was to recognize that a well-defined macroscopic thermodynamic state can be realized by several different microscopic states. The final result is that the number of microstates, nowadays called *multiplicity*, belonging to the same macrostate is proportional to the possibility for the realization of this macrostate. A closed system evolves in time into its most possible state, i. e., the one with the largest multiplicity. But in the meantime, its entropy is increasing and reaches a maximum there. Then it is obvious to say that entropy is a monotonically increasing function of multiplicity

$$S = S(\Omega_n), \tag{7.27}$$

where Ω is the number of microstates and n denotes the macrostate.

If once known, the relation (7.27) would allow us to compute the entropy simply by counting the number of microstates that make up the same macrostate n. But how to find Ω_n for a large number of particles in a systematic way? To begin with, take a very simple system with only 4 distinguishable particles or balls divided into two chambers with identical size. We define the 5 possible macrostates with $n = 0\ldots4$, corresponding to the number of balls in say the right chamber, Figure 7.4. But there exist $2^4 = 16$ microstates. To determine and count them, we number the particles with 1...4. The microstates read in detail:

microstate	left chamber	right chamber	macrostate *n*
1	1,2,3,4	–	0
2	2,3,4	1	1
3	1,3,4	2	1
4	1,2,4	3	1
5	1,2,3	4	1
6	1,2	3,4	2
7	1,3	2,4	2
8	1,4	2,3	2
9	2,3	1,2	2
10	2,4	1,3	2
11	3,4	1,2	2
12	1	2,3,4	3
13	2	1,3,4	3
14	3	1,2,4	3
15	4	1,2,3	3
16	–	1,2,3,4	4

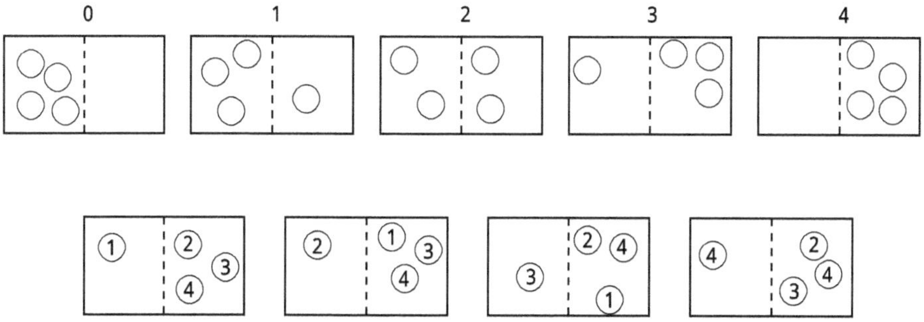

Figure 7.4: Top row: 4 particles divided into two chambers make up 5 different macrostates. Bottom row: for indistinguishable particles, all 4 microstates shown belong to the same macrostate no 3.

Thus we have

$$\Omega_0 = \Omega_4 = 1, \quad \Omega_1 = \Omega_3 = 4, \quad \Omega_2 = 6\,.$$

Throwing the balls arbitrarily in the system, the probability to find the macrostate n is given by

$$P_n = \Omega_n/16 \tag{7.28}$$

and highest for macrostate 2 (thermal equilibrium). To formulate (7.28), we have already used the fundamental theorem of statistical physics:

The microstates of a closed system are equally probable.

In our case, 16 microstates are present and each microstate has the probability 1/16. It can be seen that the general formula for Ω_n is given by the binomial coefficient "N choose n" and reads

$$\Omega_n = \frac{N!}{n!(N-n)!} = \binom{N}{n}, \tag{7.29}$$

where N is the total number of particles and n the number of particles, e. g., in the right chamber. For the probability of macrostate n one has

$$P_n = \frac{1}{2^N}\binom{N}{n}. \tag{7.30}$$

For only 4 particles the probabilities are not dramatically different. But remember, in thermodynamics N is huge ($\sim 10^{23}$), and if we take only $N = 1000$ particles, the probability to find the equilibrium state $n = 500$ or a state nearby is much higher than to find an only slightly more "sorted" state, such as $n = 550$. For $N = 1000$, one has

$$P_{500} = 0.025225, \quad P_{550} = 0.00017, \quad \frac{P_{500}}{P_{550}} = 149\,;$$

see also Figure 7.5. Expanding the binomial coefficient around its maximum $N/2$, it can be shown that the width of the probability distribution is $\sim \sqrt{N}$.

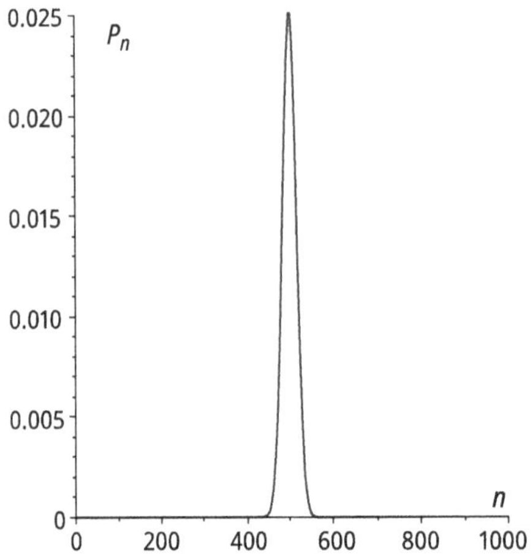

Figure 7.5: Probability for macrostate n with n particles in the right chamber, for a total of $N = 1000$ particles.

7.3.2 Entropy and multiplicity

It still remains to determine the functional dependence (7.27) and to define entropy on statistical grounds. Consider two separated isolated systems with multiplicities Ω_A, Ω_B for a certain combination of states. From there we compute the entropies

$$S_A = S(\Omega_A), \quad S_B = S(\Omega_B) .$$

The total entropy is simply

$$S_T = S_A + S_B,$$

because the entropy is extensive (additive) and the systems are isolated. What about the multiplicities? The total multiplicity is the product

$$\Omega_T = \Omega_A \cdot \Omega_B,$$

and therefore

$$S(\Omega_A \cdot \Omega_B) = S(\Omega_A) + S(\Omega_B) .$$

The only possibility to solve this equation for arbitrary Ω is given by

$$S(\Omega) = k_B \ln \Omega \tag{7.31}$$

with an arbitrary but positive constant k_B having the dimension of entropy or energy/temperature. We shall see that k_B is fixed by the chosen temperature scale. For the ideal gas temperature (Kelvin scale), it is just what is now called *Boltzmann's constant.*

7.3.3 Partitioning the phase space

The 6D phase space is a combination of 3D position space ($\vec{r}$) and 3D velocity space ($\vec{v}$). Consider N particles in a closed cube with length L and volume L^3. The macro variables can be volume, particle number, and total energy; the micro variables are position and velocity of each particle. How can we determine between different states? If two arbitrarily different $\vec{r}$ or $\vec{v}$ values would already made up two different states, we would have infinitely many of them, even for $N = 1$. Moreover, they would be uncountable. For this reason, the phase space is divided (partitioned) into small cubic cells of length $\Delta x = \Delta y = \Delta z$ and volume $\Delta V_r = \Delta x \Delta y \Delta z$; the same for the velocities. If a particle moves to another (adjacent) cell (or changes velocity due to collisions), the system accesses another state.

7.3.3.1 Position space

We begin with the position space. Instead of throwing N particles into two chambers, as in Figure 7.4, we distribute now N particles to $K = V/\Delta V_r$ cells; Figure 7.6. But there can again be more than one particle in one cell. Let n_i be the number of particles in cell i, then the generalization of equation (7.29) reads

$$\Omega(n_1 \dots n_K) = \frac{N!}{n_1!n_2!\dots n_K!}, \tag{7.32}$$

where the multiplicity depends now on the whole set of occupation numbers n_i defining the corresponding macrostate. The n_i are not all independent, because they must sum up to N:

$$\sum_i^K n_i = N. \tag{7.33}$$

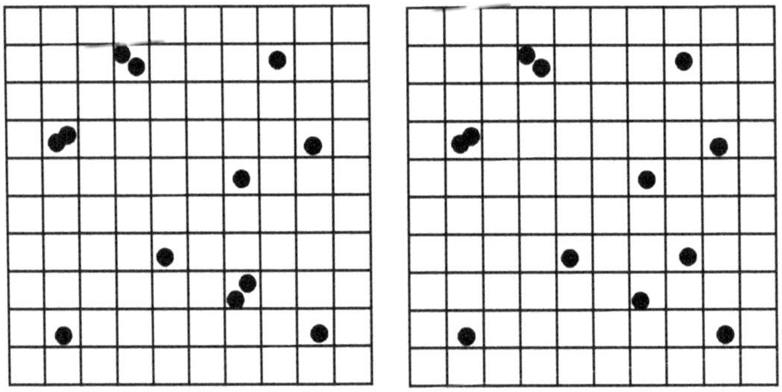

Figure 7.6: Discretization of phase space, here position space only. Two slightly different macrostates having different multiplicities of Ω = 59 875 200 (left) and Ω = 119 750 400 (right).

Inserting (7.32) into (7.31) yields for the entropy of a given macrostate,

$$S(n_1 \dots n_K) = k_B \ln\left[\frac{N!}{n_1!n_2!\dots n_K!}\right]. \tag{7.34}$$

To continue, we rely on the fact that n_i is always very large. This of course may not be the case for sufficiently rarified classical gases or for quantum gases in general. In a quantum gas of fermions, n_i can be only either zero or one, demanding for a different computation of S. But here we shall assume that Stirling's formula

$$\ln(X!) \approx X\ln(X) - X, \quad X \gg 1 \tag{7.35}$$

is valid for all n_i, and of course for N, and convert (7.34) into

$$S(n_1 \dots n_K) = -k_B \sum_i^K n_i(\ln(n_i) - \ln(N)) = -Nk_B \sum_i^K \frac{n_i}{N} \ln\left(\frac{n_i}{N}\right), \tag{7.36}$$

where we used (7.33). Next we introduce the probability to find a particle in cell i

$$P_i^r = \frac{n_i}{N} .$$

This is a value between zero and one, and clearly

$$\sum_i^K P_i^r = 1 \tag{7.37}$$

(the superscript r relates to space.) Note that NP_i^r/V is similar to the number density n defined earlier in (7.5). Then we may write (7.36) as

$$S(P_1^r \dots P_K^r) = -Nk_B \sum_i^K P_i^r \ln(P_i^r) . \tag{7.38}$$

Which set of P_i^r (or which macrostate) maximizes the entropy, or, in other words, is reached in global thermal equilibrium? Therefore, we must determine the zero of the variation of S under the constraint (7.37):

$$\delta\left[-\sum_i^K P_i^r \ln(P_i^r) + \lambda \sum_i^K P_i^r\right] = -\sum_i^K [1 + \ln(P_i^r) - \lambda]\delta P_i^r = 0$$

with the Lagrange parameter λ. The P_i^r can be varied independently, leading to

$$P_i^r = \mathrm{e}^{\lambda-1} .$$

As expected, entropy is maximized if all P_i^r have the same value, which is due to (7.37),

$$P_i^r = \frac{1}{K} = \frac{\Delta V_r}{V} . \tag{7.39}$$

The entropy for this equilibrium state is

$$S_{\max} = Nk_B \ln(K) .$$

7.3.3.2 Velocity space

For the three velocity space dimensions, the treatment is very similar. However, the three components of $\vec{v}$ run from $-\infty$ to $+\infty$, and infinitely many cells of volume ΔV_v exist. Defining n_i as the number of particles having velocity $\vec{v}_i$, i. e., laying in the i-th velocity cell, the expression

$$P_i^v = \frac{n_i}{N}$$

is the probability to find a particle with velocity $\vec{v}_i$. Then we can derive an expression for the entropy similar to (7.38), but with P_i^r replaced by P_i^v and the sum running from $-\infty$ to $+\infty$. The most important difference is that we have two constraints now: normalization of the probabilities as (7.37),

$$\sum_{i=-\infty}^{\infty} P_i^v = 1, \tag{7.40}$$

and a fixed inner total energy E, because we consider a closed system with E as a given macroscopic variable. For a perfect gas, E is kinetic energy only, and it can be written in the form

$$\frac{E}{N} = E_k = \frac{1}{2}m \sum_{i=-\infty}^{\infty} P_i^v v_i^2 = \text{const.}\,, \tag{7.41}$$

where E_k is the average kinetic energy per particle. The condition for vanishing variation with respect to P^v then reads

$$\delta\left[-\sum_{i=-\infty}^{\infty} P_i^v \ln(P_i^v) + \lambda \sum_{i=-\infty}^{\infty} P_v^r - \frac{\beta}{2} m \sum_{-\infty}^{\infty} P_i^v v_i^2\right] = 0$$

and yields

$$P_i^v = \exp\left(\lambda - 1 - \frac{\beta}{2} m v_i^2\right) \tag{7.42}$$

for the equilibrium velocity distribution. Determining the two Lagrange parameters λ, β is a bit more tricky. Inserting P_i^v in the constraint (7.41) yields

$$E_k = \frac{m}{2} e^{\lambda-1} \sum_{i=-\infty}^{\infty} v_i^2 \exp\left(-\frac{\beta}{2} m v_i^2\right) = -e^{\lambda-1} \frac{\partial Z}{\partial \beta}$$

with the abbreviation

$$Z = \sum_{i=-\infty}^{\infty} \exp\left(-\frac{\beta}{2} m v_i^2\right), \tag{7.43}$$

which is the *partition function* for a perfect gas (see textbooks on statistical thermodynamics, e. g., [1]). The first constraint (7.40) gives

$$e^{\lambda-1} = \frac{1}{Z}, \tag{7.44}$$

and therefore

$$E_k = -\frac{1}{Z}\frac{\partial Z}{\partial \beta}\,. \tag{7.45}$$

As last step, we must compute Z. If the $\vec{v}$-cells are very small, one may approximate the sum in (7.43) as integral

$$Z = \sum_{i=-\infty}^{\infty} \exp\left(-\frac{\beta}{2}mv_i^2\right) = \frac{1}{\Delta V_v}\int_{-\infty}^{\infty} d^3\vec{v}\,\exp\left(-\frac{\beta}{2}mv^2\right) = \frac{1}{\Delta V_v}\left(\frac{2\pi}{m\beta}\right)^{3/2} \tag{7.46}$$

with the volume ΔV_v in velocity space. Inserting this in (7.45) and solving for β, one finds

$$\beta = \frac{3}{2E_k} = \frac{1}{k_B T}\,,$$

where the equipartition theorem $E_k = \frac{3}{2}k_B T$ for a monatomic gas is used. Inserting β and (7.44) into (7.42) finally yields

$$P_i^v = \left(\frac{m}{2\pi k_B T}\right)^{3/2} \Delta V_v \exp\left(-\frac{mv_i^2}{2k_B T}\right) \tag{7.47}$$

for the equilibrium distribution of the velocities. Multiplying this with P_i^r from (7.39) gives exactly the same expression we already derived with a completely different reasoning in (7.19). The only difference is the 6D phase space cell volume $\Delta V_r \Delta V_v$, which is due to the fact that (7.19) is rather a density than a probability.

Although it nowhere occurs in the final results, the choice of the cell volume $\Delta V_r \Delta V_v$ seems a bit arbitrary. Boltzmann recognized clearly about 50 years before the advent of quantum mechanics that the phase space has to be partitioned. One may say he was probably the first who used quantum mechanical ideas long before they were discovered. Werner Heisenberg's uncertainty relation formulated in 1927 states that the product of momentum and position of a particle cannot be determined more accurate than $h = 6.626 \cdot 10^{-34}$ Js, an important constant in quantum mechanics and now named *Planck's constant*. This leads naturally to a smallest cell volume in phase space of

$$\Delta V_r \Delta V_v = \left(\frac{h}{m}\right)^3 .$$

7.3.4 Statistical entropy of a perfect gas

Now we can draw a link between the Maxwell–Boltzmann distribution (7.19), derived in Section 7.2.3 and the entropy. Since (7.19) is computed for a perfect monatomic gas, we expect to find the entropy from Chapter 1, equation (1.32).

We refer to equation (7.38) and replace the probability, finding a particle with the Maxwell–Boltzmann distribution $f(\vec{v})$:

$$S = -k_B \int_V d^3\vec{r} \int_{-\infty}^{\infty} d^3\vec{v}\, f(\vec{v}) \ln f(\vec{v}) \,. \tag{7.48}$$

Contrary to (7.38), we must integrate over space and velocity, because (7.19) is a density. Inserting (7.19), we find

$$f \ln f = f\left(\frac{3}{2}\ln\left(\frac{m}{2\pi k_B T}\right) + \ln\left(\frac{N}{V}\right) - \frac{mv^2}{2k_B T}\right),$$

where we put $\vec{u} = 0$ without loss of generality. Inserting this into (7.48) yields

$$\begin{aligned} S = &-k_B\left[\frac{3}{2}\ln\left(\frac{m}{2\pi k_B T}\right) + \ln\left(\frac{N}{V}\right)\right] \underbrace{\int_V d^3\vec{r} \int_{-\infty}^{\infty} d^3\vec{v}\, f(\vec{v})}_{=N} \\ &+ \frac{m}{2T} \underbrace{\int_V d^3\vec{r} \int_{-\infty}^{\infty} d^3\vec{v}\, v^2 f(\vec{v})}_{-3k_B TN/m} \\ = &-Nk_B\left(\frac{3}{2}\ln\left(\frac{m}{2\pi k_B T}\right) + \ln\left(\frac{N}{V}\right) - \frac{3}{2}\right), \end{aligned} \tag{7.49}$$

where we used (7.5) for the first and (7.13) for the second integral. If we convert

$$\frac{3}{2}\ln\left(\frac{m}{2\pi k_B T}\right) + \ln\left(\frac{N}{V}\right) = -\frac{3}{2}\ln\left(\frac{T}{T_0}\right) - \ln\left(\frac{V}{V_0}\right) + \Phi\,,$$

where the terms confined in Φ do not depend on T and V, we finally get

$$S(V,T) = k_B N \ln\left(\frac{V}{V_0}\right) + \frac{3}{2}k_B N \ln\left(\frac{T}{T_0}\right) + S_0\,, \tag{7.50}$$

where S_0 gathers all constants left. With $R = Nk_B$ and $c_V = \frac{3}{2}Nk_B$; this is similar to (1.32).

In summary, we see that the Maxwell–Boltzmann distribution for microscopic particle velocities delivers the correct function for the entropy formulated in macroscopic variables.

The expression $f \ln f$ was originally used by Boltzmann to prove his famous *H-Theorem*. He showed that for an arbitrary non-stationary distribution out of equilibrium the functional

$$H(t) = \int_V d^3\vec{r} \int_{-\infty}^{\infty} d^3\vec{v}\, f(\vec{v},\vec{r},t) \ln f(\vec{v},\vec{r},t) \tag{7.51}$$

is always decreasing in time as long as f changes. If the Maxwellian is reached, H is at minimum, and the gas is in equilibrium. Because $S \sim -H$, this will turn out to be nothing else than the second law of thermodynamics formulated with microscopic quantities. However, for the proof, one needs to know how $f(t)$ evolves in time; this is addressed in the next chapter.

7.4 Estimation of transport coefficients

So far, we considered the gas in global equilibrium with a homogeneous density and a Gaussian velocity distribution. Now we assume local equilibrium and take temperature, and density space, and later, time dependent. The condition that the local equilibrium approach works is that the macroscopic variables (temperature, etc.) change on a much larger scale than the *mean free path* Λ of the microscopic particles. This can be formulated for instance for temperature:

$$\frac{|\nabla T|}{T} \ll \frac{1}{\Lambda}, \tag{7.52}$$

but also for density, pressure, mean velocity, etc.

To simplify the derivation, the computations here and in the following section contain some approximations that are not always exact. This is why the numerical factors that will turn out should not be taken too seriously. For more details see [13].

7.4.1 Mean free path, mean free time, and mobility

Assume N particles in a volume V in isotropic thermal random motion, all with the same mean speed $\langle v^2 \rangle \sim T$. Each particle has the radius R and mass m.

7.4.1.1 Mean free time

How long in average can a particle move before it collides with another one? Let us take a certain particle and follow it, starting from its last collision. If it moves with random velocity $\vec{\bar{v}}$ relatively to the others, it may hit in a certain time τ those particles that are in a cylinder with volume $\sigma|\bar{v}|\tau$, where σ is the total scattering cross section of the particle ($4\pi R^2$ for a hart sphere ball, see Chapter 8). If the particle density is $n = N/V$, then the number of other particles it meets is estimated as

$$N_0(\tau) = n\sigma|\bar{v}|\tau\,.$$

The mean free time is defined as time between two collisions. So putting $N_0 = 1$ yields

$$\tau = \frac{1}{n\sigma|\bar{v}|}\,. \tag{7.53}$$

$\bar{v}$ is the averaged relative velocity with respect to the particle, which is met first:

$$\langle \bar{v}^2 \rangle = \langle (\vec{v} - \vec{v}_1)^2 \rangle = \langle v^2 \rangle + \langle v_1^2 \rangle - 2\langle v v_1 \cos\varphi \rangle ,$$

where $\langle v_1^2 \rangle = \langle v^2 \rangle$ and φ is the angle between the two particle trajectories. We may average over the angle, because the particles move isotropic,

$$\langle \cos\varphi \rangle = 0$$

and find

$$\bar{v}^2 = \frac{2}{n} \langle v^2 \rangle = \frac{6k_BT}{m} ,$$

where we used (7.13). Then the mean free time is estimated as

$$\tau = \frac{1}{\sqrt{2} n\sigma |v|} \sim \frac{1}{\sqrt{T}} . \tag{7.54}$$

7.4.1.2 Mean free path

Another important quantity is the mean free path, the distance a particle travels in average between two collisions. This is simply given by

$$\Lambda = |v|\tau$$

and inserting (7.54), we find

$$\Lambda = \frac{1}{\sqrt{2} n\sigma} . \tag{7.55}$$

This is an important relation, which says that the mean free path depends only on the effective size of the particles and on the particle density. Temperature and particle mass plays no role!

7.4.1.3 Mobility

The last important quantity we determine is the mobility. Consider a single ion in an otherwise neutral gas. How fast is this ion moving if a certain constant electric field is applied? If we assume that the particle is free between two collisions it will accelerate to

$$mv = F\tau ,$$

where $F = eE$ is the force coming from the electric field. Then it collides and will in average lose its velocity again. So the mean velocity will be estimated as $v = F\tau/m$. The mobility μ is defined as

$$v = \mu F ,$$

and it follows that

$$\mu = \frac{\tau}{m} . \tag{7.56}$$

Since $\tau \sim 1/\sqrt{T}$ the same holds for μ. If the temperature is larger, the mobility decreases and vice versa.

7.4.2 Self diffusion

We shall estimate the diffusion coefficient describing self diffusion. Consider a homogeneously distributed particle gas, in which a few tracer particles with the same physical properties are dissolved (some green balls in a bunch of black ones); see Figure 7.7. If the tracers are inhomogeneously distributed in the beginning and all particles are in thermal random motion, they will diffuse into the black particles, ending up in a homogeneous mixture of black and green particles.

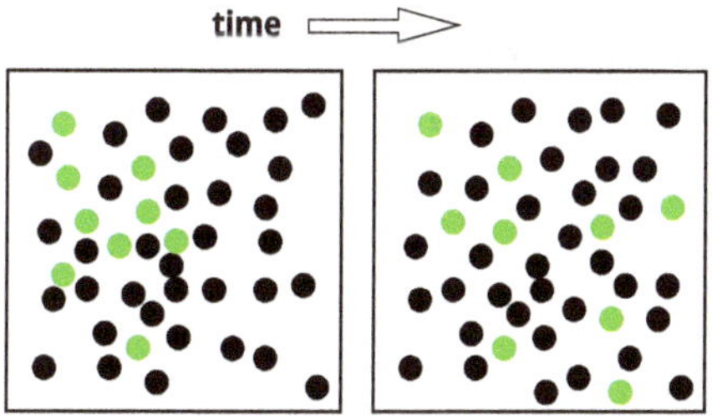

Figure 7.7: Green tracer particles in homogeneously distributed black particles. In the beginning (left frame), the system is out of equilibrium. After a certain time, equilibrium is reached due to (self) diffusion.

Phenomenologically this situation was studied in Section 4.3, and a diffusion equation (4.46) was derived. Here we shall determine the diffusion coefficient D on statistical grounds.

Clearly, the initial situation shown in the left frame of Figure 7.7 is not in equilibrium. If the system is isothermal, only the particle density of the tracers becomes a function of space and time. Now consider Figure 7.8. We wish to compute the number of particles traveling through the plane $z = 0$ with the surface A in a time interval δt. A particle having the z-velocity $-v_z$ is at δt at the surface when it was at $t = 0$ at $v_z \delta t$. But the particles have random velocities, and we must take the mean value of v_z over all negative velocities. Let $f(\vec{r}, \vec{v}, t)$ be the PDF of the particles according to (7.20). Then

$$\begin{aligned}\delta N_I &= -n_I A \delta t \left(\frac{m}{2\pi k_B T} \right)^{3/2} \int_{-\infty}^{\infty} dv_x \int_{-\infty}^{\infty} dv_y \int_{-\infty}^{0} dv_z v_z e^{-\frac{m}{2k_B T} v^2} \\ &= n_I A \delta t \left(\frac{k_B T}{2\pi m} \right)^{1/2}\end{aligned}$$

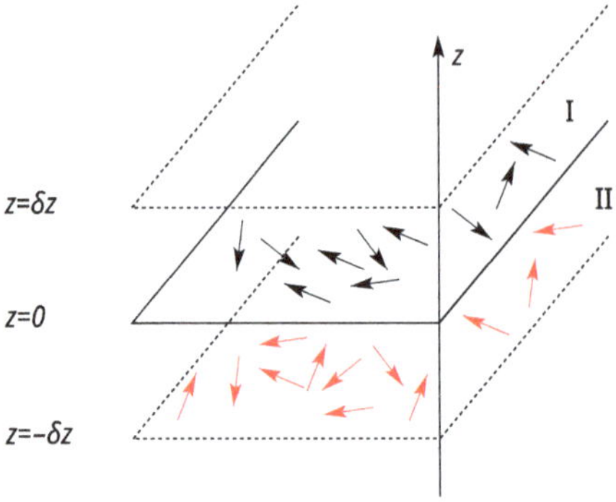

Figure 7.8: Thermal motion below and above a horizontal plate at $z = 0$. If the density at $z > 0$ is smaller than that for $z < 0$, more particles travel in average from II to I, leading to a diffusive flux $\sim -dn/dx$ (see text).

is the number of particles traveling downwards (from I to II) in the time interval δt. Here, n_I denotes the particle density in the upper half, $z > 0$. The same computation, but now with mean over positive v_z, yields for the number of particles traveling upwards from II to I:

$$\delta N_{II} = n_{II} A\delta t \left(\frac{m}{2\pi k_B T}\right)^{3/2} \int_{-\infty}^{\infty} dv_x \int_{-\infty}^{\infty} dv_y \int_0^{\infty} dv_z v_z e^{-\frac{m}{2k_B T}v^2}$$
$$= n_{II} A\delta t \left(\frac{k_B T}{2\pi m}\right)^{1/2} .$$

The mass flux (particles per time and area) is then given by

$$j_z = \frac{\delta N_{II} - \delta N_I}{A\delta t} = (n_{II} - n_I)\left(\frac{k_B T}{2\pi m}\right)^{1/2} . \tag{7.57}$$

Now we come to the crucial point: since $n = n(\vec{r})$, where should we determine n_I, n_{II}? It should be rather close to the plane $z = 0$, say somewhere at $z = \pm\delta z$, where δz is on the microscopic length scale. Then we may write

$$n_{II} - n_I = \frac{n(z - \delta z) - n(z + \delta z)}{2\delta z} 2\delta z = -2\delta z \frac{dn}{dz} , \tag{7.58}$$

where the last equation holds if condition (7.52) is fulfilled.

On the other hand, the diffusion equation (see Section 4.3) reads

$$\partial_t n = -\nabla \cdot \vec{j} = D\nabla^2 n ,$$

and thus

$$\vec{j} = -D\nabla n . \tag{7.59}$$

Inserting (7.58) into (7.57), we readily may generalize it to three dimensions and write

$$\vec{j} = -2\delta z \left(\frac{k_B T}{2\pi m}\right)^{1/2} \nabla n,$$

and with (7.59), we find for the diffusion coefficient

$$D = 2\delta z \left(\frac{k_B T}{2\pi m}\right)^{1/2}.$$

It is reasonable to regard δz equal to or at least of the same size as the mean free path Λ of the gas. A more detailed computation yields

$$\delta z = 2\Lambda/3$$

(see [13]), and finally

$$D = \frac{\Lambda}{3}\left(\frac{8k_B T}{\pi m}\right)^{1/2}. \tag{7.60}$$

Expressing Λ with (7.55), we can write D as function of particle properties, temperature and density

$$D = \frac{2}{3n\sigma}\left(\frac{k_B T}{\pi m}\right)^{1/2}. \tag{7.61}$$

An important note is that $D \sim \sqrt{T}$ and $D \sim 1/n$; both are experimentally well confirmed. Finally, we can express Λ with the mobility $\Lambda = \mu m v$ and get

$$D = \frac{8}{3\pi}\mu k_B T\,. \tag{7.62}$$

Here, the formula with the correct numerical coefficient would read $D = \mu k_B T$; see [14].

7.4.3 Thermal conduction

The estimation for the thermal conductivity introduced in (2.68) is similar. The only difference is that instead of the tracer particle flux, we now must compute the thermal energy flux in δt through the surface A at $z = 0$. For that, we assume different temperatures in I and II, but a homogeneous particle density $n_I = n_{II} = n$. For the energy flowing downwards, we have

$$\begin{aligned}\delta E_I &= -\frac{1}{2} mnA\delta t\left(\frac{m}{2\pi k_B T_I}\right)^{3/2} \int_{-\infty}^{\infty} dv_x \int_{-\infty}^{\infty} dv_y \int_{-\infty}^{0} dv_z v^2 v_z e^{-\frac{m}{2k_B T_I}v^2} \\ &= nA\delta t\left(\frac{2}{\pi m}\right)^{1/2} (k_B T_I)^{3/2}\,.\end{aligned}$$

The energy flowing upwards is found as

$$\delta E_{II} = \frac{1}{2} mnA\delta t \left(\frac{m}{2\pi k_B T_{II}}\right)^{3/2} \int_{-\infty}^{\infty} dv_x \int_{-\infty}^{\infty} dv_y \int_{0}^{\infty} dv_z v^2 v_z e^{-\frac{m}{2k_B T_{II}} v^2}$$
$$= nA\delta t \left(\frac{2}{\pi m}\right)^{1/2} (k_B T_{II})^{3/2} .$$

For the heat flux the difference reads as follows:

$$q_z = \frac{\delta E_{II} - \delta E_I}{A\delta t} = n\left(\frac{2}{\pi m}\right)^{1/2} \left((k_B T_{II})^{3/2} - (k_B T_I)^{3/2}\right) . \tag{7.63}$$

We expand the last bracket according to

$$T_I^{3/2} = \left(T + \frac{dT}{dz}\delta z\right)^{3/2} = T^{3/2} + \frac{3}{2}\sqrt{T}\frac{dT}{dz}\delta z,$$
$$T_{II}^{3/2} = \left(T - \frac{dT}{dz}\delta z\right)^{3/2} = T^{3/2} - \frac{3}{2}\sqrt{T}\frac{dT}{dz}\delta z$$

with T being the temperature at $z = 0$. From there, with (7.63),

$$q_z = -3nk_B \left(\frac{2k_B T}{\pi m}\right)^{1/2} \delta z \frac{dT}{dz} . \tag{7.64}$$

Again we put $\delta z = 2\Lambda/3$, and finally find for the thermal conductivity

$$\lambda = nk_B \left(\frac{8k_B T}{\pi m}\right)^{1/2} \Lambda . \tag{7.65}$$

As above, this can be expressed in “molecular quantities” as

$$\lambda = \frac{2k_B}{\sigma}\left(\frac{k_B T}{\pi m}\right)^{1/2} . \tag{7.66}$$

The experimentally confirmed behavior $\lambda \sim \sqrt{T}$ is found, as well as the independence from density.

7.4.4 Lewis number

In Section 6.3.3, we introduced the dimensionless Lewis number as the ratio between mass and thermal diffusion, $L = D/\kappa$. The thermal diffusivity can be expressed as

$$\kappa = \frac{\lambda}{c\rho} ,$$

with c as specific heat per molar mass M. Now we can compute L with (7.61) and (7.65) and find the universal value of

$$L = \frac{Dc\rho}{\lambda} = \frac{1}{2}$$

for monatomic gases with $c = \frac{3}{2}R/M$.

8 The Boltzmann equation

In the previous chapter, we demonstrated that a many-particle system can be statistically described by the probability distribution function (PDF) $f(\vec{v})$ with particle velocity $\vec{v}$. This concept is naturally extended to local equilibrium and $f(\vec{r}, \vec{v}, t)$. We saw that the Maxwellian (7.19) leads for the case of a closed system to maximal entropy, and can therefore be considered as (local) equilibrium distribution. Now we may ask how an arbitrary distribution even far from local equilibrium evolves in space and time. In other words, we are looking for a field equation for f. Since we have seven independent variables, this equation must be a seven-dimensional (or better 6 + 1) partial differential equation.

8.1 Kinetic equation

8.1.1 Non-interacting particles

We repeat the definition of the PDF introduced in (7.3)

$$dN = f(\vec{r}, \vec{v}, t)\; d^3\vec{r}\; d^3\vec{v}\,, \tag{8.1}$$

where dN is the number of particles in the infinitesimal volume $d^3\vec{r}\; d^3\vec{v}$. If we are looking at the same particles at an (infinitesimally) short time δt later, this volume element deforms to $d^3\vec{r}'\; d^3\vec{v}'$ and moves to $\vec{r}', \vec{v}'$. The number of particles reads now

$$dN' = f(\vec{r}', \vec{v}', t + \delta t)\; d^3\vec{r}'\; d^3\vec{v}'\,. \tag{8.2}$$

If the particle number is conserved, one must have $dN = dN'$. The new positions and velocities are found from

$$\vec{r}' = \vec{r} + \vec{v}\delta t + O(\delta t^2), \quad \vec{v}' = \vec{v} + \vec{a}\delta t + O(\delta t^2)\,, \tag{8.3}$$

where the acceleration $\vec{a}$ is due to external forces (gravity, electric or magnetic fields for charged particles). The volume elements transform according to

$$d^3\vec{r}'\; d^3\vec{v}' = \det \underline{J}\; d^3\vec{r}\; d^3\vec{v}$$

with the 6×6 Jacobian

$$\underline{J} = \frac{\partial(\vec{r}', \vec{v}')}{\partial(\vec{r}, \vec{v})} = \begin{pmatrix} \dfrac{\partial r_i'}{\partial r_j} & \dfrac{\partial v_i'}{\partial r_j} \\[2ex] \dfrac{\partial r_i'}{\partial v_j} & \dfrac{\partial v_i'}{\partial v_j} \end{pmatrix}. \tag{8.4}$$

https://doi.org/10.1515/9783110794137-010

Inserting (8.3), one obtains

$$\underline{J} = \begin{pmatrix} 1 & 0 & 0 & \frac{\partial a_1}{\partial r_1}\delta t & \frac{\partial a_1}{\partial r_2}\delta t & \frac{\partial a_1}{\partial r_3}\delta t \\ 0 & 1 & 0 & \frac{\partial a_2}{\partial r_1}\delta t & \frac{\partial a_2}{\partial r_2}\delta t & \frac{\partial a_2}{\partial r_3}\delta t \\ 0 & 0 & 1 & \frac{\partial a_3}{\partial r_1}\delta t & \frac{\partial a_3}{\partial r_2}\delta t & \frac{\partial a_3}{\partial r_3}\delta t \\ \delta t & 0 & 0 & 1+\frac{\partial a_1}{\partial v_1}\delta t & \frac{\partial a_1}{\partial v_2}\delta t & \frac{\partial a_1}{\partial v_3}\delta t \\ 0 & \delta t & 0 & \frac{\partial a_2}{\partial v_1}\delta t & 1+\frac{\partial a_2}{\partial v_2}\delta t & \frac{\partial a_2}{\partial v_3}\delta t \\ 0 & 0 & \delta t & \frac{\partial a_3}{\partial v_1}\delta t & \frac{\partial a_3}{\partial v_2}\delta t & 1+\frac{\partial a_3}{\partial v_3}\delta t \end{pmatrix}, \qquad (8.5)$$

and from there

$$\det \underline{J} = 1 + \sum_{i=1}^{3} \frac{\partial a_i}{\partial v_i}\,\delta t + O(\delta t^2)\,. \qquad (8.6)$$

Here we assume the not very restrictive property

$$\sum_{i=1}^{3} \frac{\partial a_i}{\partial v_i} = 0,$$

which is valid for most of the external forces (also for the Lorentz force). Thus we may write

$$d^3\vec{r}'\,d^3\vec{v}' = d^3\vec{r}\,d^3\vec{v} + O(\delta t^2)\,. \qquad (8.7)$$

8.1.2 Collisions

If there is no particle interaction, the number dN' from (8.2) must be equal to dN from (8.1) (particle conservation). But if the particles move completely independently from each other, thermalization would not be possible and (local) equilibrium would never be reached, even for a closed system. Thus we must at least allow for binary collisions. If two particles meet at some position $\vec{r}$, they will still be at the same place right after collision, but their velocities will change. Thus, a particle can disappear from the volume element $d^3\vec{v}$ and show up in another element than $d^3\vec{v}'$, leading to a difference in dN and dN':

$$dN' - dN = dN_c\,. \qquad (8.8)$$

If $dN_c < 0$, this is the number of particles originally in $d^3\vec{v}$ that do not arrive in $d^3\vec{v}'$. If, on the other hand, $dN_c > 0$, particles are scattered into $d^3\vec{v}'$ from somewhere else.

8.1.3 Kinetic equation

Finally, we may expand (8.2) in a Taylor series up to first order in δt and obtain

$$dN' = \left\{ f(\vec{r}, \vec{v}, t) + \left[\frac{\partial f}{\partial t} + \sum_{i=1}^{3} \frac{\partial f}{\partial v_i} a_i + \sum_{i=1}^{3} \frac{\partial f}{\partial r_i} v_i \right] \delta t \right\} d^3\vec{r}'\, d^3\vec{v}' . \tag{8.9}$$

Inserting this into (8.8) and using (8.1) and (8.7) yields the *kinetic equation*

$$\boxed{\frac{\partial f}{\partial t} + \sum_{i=1}^{3} a_i \frac{\partial f}{\partial v_i} + \sum_{i=1}^{3} v_i \frac{\partial f}{\partial r_i} = \frac{\delta f}{\delta t}} , \tag{8.10}$$

where we have substituted

$$dN_c = \delta f\, d^3\vec{r}\, d^3\vec{v} . \tag{8.11}$$

Here, δf denotes the change of the PDF f due to collisions, and $\delta f/\delta t$ is a formal writing of the so-called collision term, giving the rate of change of the PDF due to collisions. We shall examine the collision term that contains the whole physics of the collision in more detail below. Once the collision term is known, we have reached our goal to find a PDE for the evolution of f under internal (collisions) and external ($\vec{a}$) forces.

8.2 Boltzmann equation

To eventually formulate the Boltzmann equation, the collision term is expressed as nonlinear functional of the PDF, which we shall achieve next.

8.2.1 Boltzmann collision integral

The collision term in (8.10) denotes the rate with which the number of particles in the interval $d^3\vec{r}\; d^3\vec{v}$ changes due to collisions. This can be divided into two non-negative parts,

$$\frac{\delta f}{\delta t} = \left(\frac{\delta f}{\delta t}\right)_+ - \left(\frac{\delta f}{\delta t}\right)_- , \tag{8.12}$$

where the first part denotes the number of particles that are scattered into $d^3\vec{r}\; d^3\vec{v}$ from all other places in phase space (replenishment); the second part the number that is scattered from $d^3\vec{r}\; d^3\vec{v}$ into other parts (depletion).

We assume a dilute gas and neglect others than binary collisions. Consider Figure 8.1. Let

$$W(\vec{v}', \vec{v}_1'; \vec{v}, \vec{v}_1) \tag{8.13}$$

be the probability rate (at $\vec{r}$), which two incoming particles with $\vec{v}$, $\vec{v}_1$ leave the collision with the velocities $\vec{v}'$, $\vec{v}_1'$. If there exist certain conservation laws, such as momentum and energy conservation, they must be incorporated into W. We may formulate the depletion part as

$$\left(\frac{\delta f}{\delta t}\right)_- = \int d^3\vec{v}_1 d^3\vec{v}' d^3\vec{v}_1'\; W(\vec{v}', \vec{v}_1'; \vec{v}, \vec{v}_1) f(\vec{r}, \vec{v}, t) f(\vec{r}, \vec{v}_1, t), \tag{8.14}$$

where all integrals in this section run from $-\infty$ to ∞. The same can be done for the replenishment part

$$\left(\frac{\delta f}{\delta t}\right)_+ = \int d^3\vec{v}_1 d^3\vec{v}' d^3\vec{v}_1'\; W(\vec{v}, \vec{v}_1; \vec{v}', \vec{v}_1') f(\vec{r}, \vec{v}', t) f(\vec{r}, \vec{v}_1', t)\,. \tag{8.15}$$

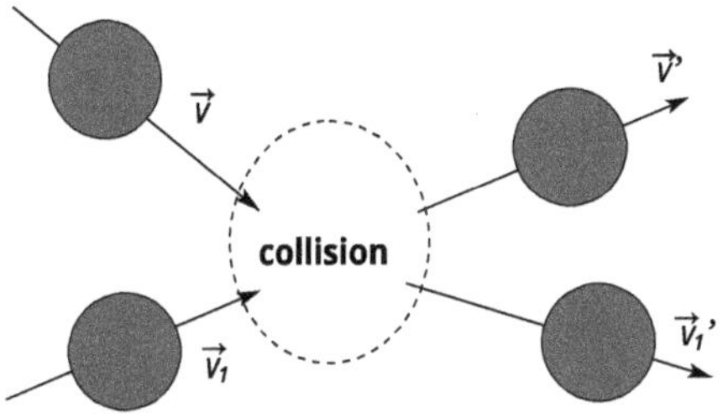

Figure 8.1: Binary collision of two equal particles. In an elastic collision, momentum and energy are conserved.

If the binary collision is elastic, the microscopic world is reversible and each collision can be performed backwards with the same transition rate, therefore

$$W(\vec{v}', \vec{v}_1'; \vec{v}, \vec{v}_1) = W(\vec{v}, \vec{v}_1; \vec{v}', \vec{v}_1') \tag{8.16}$$

must hold. Thus we have the collision integral from (8.12),

$$\left(\frac{\delta f}{\delta t}\right) = \int d^3\vec{v}_1 d^3\vec{v}' d^3\vec{v}_1'\; W(\vec{v}, \vec{v}_1; \vec{v}', \vec{v}_1')(f' f_1' - f f_1) \tag{8.17}$$

with the abbreviations

$$f = f(\vec{r}, \vec{v}, t), \quad f' = f(\vec{r}, \vec{v}', t), \quad f_1 = f(\vec{r}, \vec{v}_1, t), \quad f_1' = f(\vec{r}, \vec{v}_1', t)\,.$$

8.2.2 Boltzmann equation

Inserting (8.17) into (8.10), we finally arrive at the *Boltzmann equation*

$$\boxed{\frac{\partial f}{\partial t} + \sum_{i=1}^{3} a_i \frac{\partial f}{\partial v_i} + \sum_{i=1}^{3} v_i \frac{\partial f}{\partial r_i} = \int d^3\vec{v}_1 d^3\vec{v}' d^3\vec{v}_1' \; W(\vec{v},\vec{v}_1;\vec{v}',\vec{v}_1')(f'f_1' - f f_1)} \; . \tag{8.18}$$

This equation, written down first by Ludwig Boltzmann in 1872, is a nonlinear integro-differential equation for the PDF. Boundary and initial condition given, one could in principle, compute $f(\vec{r},\vec{v},t)$ if the collision physics included in W is known. With the Boltzmann equation, one has reduced the original problem stated in (7.1) from $6N+1$ to only $6+1$ dimensions. Nevertheless, solving (8.18) exactly is still impossible, and one has to devise some approximations and tricks, as we shall see in what follow.

Sometimes it is more convenient to formulate the PDF with random velocities $\vec{w} = \vec{v} - \vec{u}$:

$$f = f(\vec{r},\vec{w},t) \, .$$

It is important to note that contrary to $\vec{v}$, $\vec{w}$ depends on $\vec{r}$ and t. Thus one has to replace

$$\frac{\partial f}{\partial t} \longrightarrow \frac{\partial f(\vec{r},\vec{w},t)}{\partial t} + \sum_i^3 \frac{\partial f(\vec{r},\vec{w},t)}{\partial w_i}\frac{\partial w_i}{\partial t} = \frac{\partial f(\vec{r},\vec{w},t)}{\partial t} - \sum_i^3 \frac{\partial f(\vec{r},\vec{w},t)}{\partial w_i}\frac{\partial u_i}{\partial t} \, ,$$

$$\frac{\partial f}{\partial r_j} \longrightarrow \frac{\partial f(\vec{r},\vec{w},t)}{\partial r_j} + \sum_i^3 \frac{\partial f(\vec{r},\vec{w},t)}{\partial w_i}\frac{\partial w_i}{\partial r_j} = \frac{\partial f(\vec{r},\vec{w},t)}{\partial r_j} - \sum_i^3 \frac{\partial f(\vec{r},\vec{w},t)}{\partial w_i}\frac{\partial u_i}{\partial r_j}$$

and obtains the Boltzmann equation in the form

$$\frac{\partial f}{\partial t} + \sum_{i=1}^{3}\left[a_i - \frac{\partial u_i}{\partial t} - \sum_{j=1}^{3}(u_j + w_j)\frac{\partial u_i}{\partial r_j}\right]\frac{\partial f}{\partial w_i} + \sum_{i=1}^{3}(u_i + w_i)\frac{\partial f}{\partial r_i} = \left(\frac{\delta f}{\delta t}\right) . \tag{8.19}$$

The collision integral is the same as (8.18), but $\vec{v}$ replaced by $\vec{w}$, and so on.

If a multi-component gas mixture of K components is considered, as in Section 7.2.5, the PDFs for each component are ruled by a set of K Boltzmann equations coupled by the collision integrals. The same derivation as for (8.18) leads now to

$$\frac{\partial f^k}{\partial t} + \sum_{i=1}^{3} a_i \frac{\partial f^k}{\partial v_i} + \sum_{i=1}^{3} v_i \frac{\partial f^k}{\partial r_i} = \sum_{\ell}^{K} \int d^3\vec{v}_1 d^3\vec{v}' d^3\vec{v}_1' \; W_{k\ell}(\vec{v},\vec{v}_1;\vec{v}',\vec{v}_1')((f^k)'(f_1^\ell)' - f^k f_1^\ell) \, .$$

8.2.3 Collision integral and conservation laws

For an elastic collision, momentum and energy are conserved, leading to

$$\vec{v} + \vec{v}_1 = \vec{v}\,' + \vec{v}_1', \quad v^2 + v_1^2 = v'^2 + v_1'^2 \,. \tag{8.20}$$

From there, it is clear that if f is a local Maxwell–Boltzmann PDF as (7.20), $f f_1 = f' f_1'$ holds, and the collision integral (8.17) is always zero.

With (8.20), the 9-fold integral in (8.17) can be reduced to a 5-fold integral. We perform the computation in relative and mass centered coordinates (Figure 8.2),

$$\vec{v}_r = \vec{v} - \vec{v}_1, \quad \vec{v}_c = \frac{\vec{v} + \vec{v}_1}{2} \tag{8.21}$$

and the same for the primed variables. Note that due to (8.20)

$$\vec{v}_c = \vec{v}_c', \quad |v_r| = |v_r'| \,. \tag{8.22}$$

The determinant of the Jacobian $\partial(\vec{v}, \vec{v}_1)/\partial(\vec{v}_c, \vec{v}_r)$ is equal to one, and therefore $d^3\vec{v}\,' d^3\vec{v}_1' = d^3\vec{v}_r' d^3\vec{v}_c'$. Then, (8.17) can be written as

$$\left(\frac{\delta f}{\delta t}\right) = \int d^3\vec{v}_1 d^3\vec{v}_r' d^3\vec{v}_c' \; \tilde{W}(v_r, \vartheta')\delta(v_r - v_r')\delta(\vec{v}_c - \vec{v}_c')(f' f_1' - f f_1) \,, \tag{8.23}$$

where ϑ' is the angle between v_1 and v_1', the scattering angle, and the δ-functions account for the conservations (8.22). We used

$$\begin{aligned} W(\vec{v}, \vec{v}_1; \vec{v}\,', \vec{v}_1') &= W(\vec{v}_c + \vec{v}_r/2, \vec{v}_c - \vec{v}_r/2, \vec{v}_c' + \vec{v}_r'/2, \vec{v}_c' - \vec{v}_r'/2) \\ &= \tilde{W}(v_r, \vartheta')\delta(v_c - v_c')\delta(\vec{v}_r - \vec{v}_r') \,. \end{aligned}$$

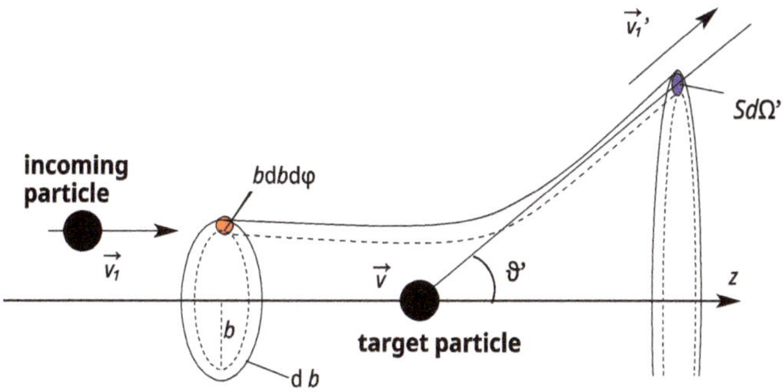

Figure 8.2: An incoming particle with velocity $\vec{v}_1$ is scattered by a target particle with $\vec{v}$. The impact parameter b is the closest distance the particles would have if they would not interact. The small read area is related to the blue one by the differential cross section $S(\varphi', \vartheta')$.

Here, $\tilde{W}$ can only depend on the absolute value of $\vec{v}_r$ and the angle ϑ' between $\vec{v}_1$ and $\vec{v}_1'$. The reason for this is very fundamental: isotropy of space (the whole process in Figure 8.2 can be rotated by arbitrary angles) and Galilean invariance. Introducing spherical coordinates v_r', φ', ϑ', we can perform the integration over $d^3\vec{v}_c'$ and dv_r' and obtain from (8.23)

$$\left(\frac{\delta f}{\delta t}\right) = \int d^3\vec{v}_1 d^2\Omega'\ v_r^2\ \tilde{W}(v_r, \vartheta')(f'f_1' - f f_1) \tag{8.24}$$

with the solid angle element $d^2\Omega' = \sin\vartheta' d\vartheta' d\varphi'$.

The transition rate $\tilde{W}$ depends on the physics of the collision. Consider Figure 8.2. The differential flux density of the incoming beam having the velocity $\vec{v}_1$ is given as

$$dJ = |v_r| f(\vec{v}_1) d^3\vec{v}_1\,. \tag{8.25}$$

Let the z-axis of the system point in direction of $\vec{v}_r$, $\vec{v}_r = |v_r|\hat{e}_z$. The number of particles per time scattered into the solid angle element $d\Omega'$ at a certain angle ϑ', φ' is directly related to the number of incoming particles per time through the element $b\,db\,d\varphi$, where b is the impact parameter:

$$dN_{\Omega'} = b\,db\,d\varphi\ dJ = S(\varphi', \vartheta')d\Omega' dJ = S(\varphi', \vartheta')\sin\vartheta' d\vartheta' d\varphi' dJ\,. \tag{8.26}$$

Here we have introduced the *differential cross section* as

$$S(\varphi', \vartheta') = \frac{b\,db\,d\varphi}{d\Omega'} = \frac{b\,db\,d\varphi}{\sin\vartheta' d\vartheta' d\varphi'} = \frac{b}{\sin\vartheta'}\frac{db}{d\vartheta'}, \tag{8.27}$$

which has the dimension of an area and contains the physics of the scattering process. For a rotational symmetric pair potential, one has $S = S(\vartheta')$ only. Note that $dN_{\Omega'}$ is equal to the number of particles that leave the phase space at $\vec{v}\ d^3\vec{v}$, accounting for depletion in the collision integral. With (8.26) and (8.25), we can compute the depletion part of the collision integral (8.14) after integrating over all angles and all particle velocities $\vec{v}_1$ as

$$\left(\frac{\delta f}{\delta t}\right)_- = \int d^3\vec{v}_1 d^2\Omega'\ |\vec{v} - \vec{v}_1|\ S(\vartheta') f(\vec{v}) f(\vec{v}_1)\,. \tag{8.28}$$

Comparing this expression with the depletion part of (8.24), we find the relation

$$\tilde{W}(v_r, \vartheta') = \frac{S(\vartheta')}{v_r}\,. \tag{8.29}$$

The functional form of S depends on the physics of particle interaction. The simplest case is a hard-core repulsion of spherical particles, reflecting the interactions in a perfect monatomic gas. Then one can easily compute a relation between the scattering angle ϑ' and the impact parameter b as

$$b = 2R\cos\left(\frac{\vartheta'}{2}\right), \tag{8.30}$$

where R is the particle radius (see Figure 8.3). Inserting this into (8.27) yields for the differential cross section $S = R^2$. Finally, the collision integral for hard spheres (perfect gas) reads

$$\left(\frac{\delta f}{\delta t}\right) = R^2 \int d^3\vec{v}_1 d^2\Omega' \, |\vec{v} - \vec{v}_1| \, (f' f_1' - f f_1) \,. \tag{8.31}$$

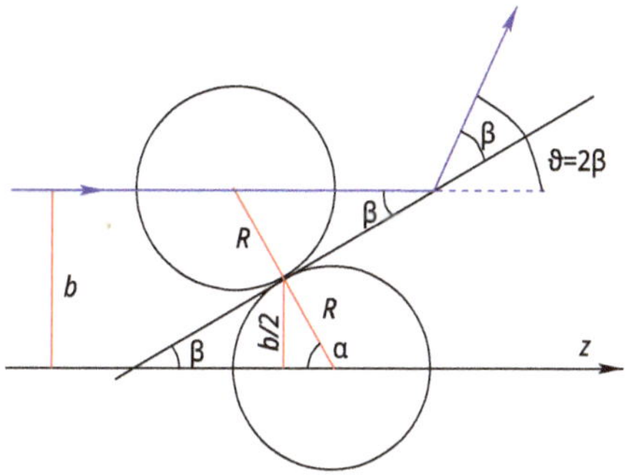

Figure 8.3: Scattering process of two hard spheres with $\frac{b}{2} = R\sin\alpha$; $\alpha + \beta = \pi/2$ follows the relation (8.30).

From (8.28), it is straightforward to find the collision integral for other pair potentials. For a repelling Coulomb potential

$$U(r) = \frac{\alpha}{r}, \quad \alpha > 0,$$

one finds [15]

$$b(\vartheta) = \frac{2\alpha}{mv^2}\cot\left(\frac{\vartheta}{2}\right)$$

with m as the particle mass and v as the relative velocity for $r \to \infty$. Inserting this into (8.27) yields Rutherford's formula

$$S(\vartheta) = \frac{1}{\sin^4(\frac{\vartheta}{2})}\left(\frac{\alpha}{mv^2}\right)^2,$$

and for the collision integral

$$\left(\frac{\delta f}{\delta t}\right) = \left(\frac{\alpha}{m}\right)^2 \int d^3\vec{v}_1 d^2\Omega' \, \frac{1}{\sin^4(\frac{\vartheta'}{2})} \, \frac{1}{|\vec{v} - \vec{v}_1|^3} \, (f' f_1' - f f_1) \,. \tag{8.32}$$

8.2.4 H-Theorem

Boltzmann's H-functional was already introduced in Section 7.3.4, equation (7.51). We repeat it here:

$$H(t) = \int_V d^3\vec{r} \int_{-\infty}^{\infty} d^3\vec{v}\, f(\vec{v},\vec{r},t) \ln f(\vec{v},\vec{r},t)\,. \tag{8.33}$$

We saw that $H \sim -S$ where S is the entropy. Consequently, H should take its minimum for f being the equilibrium Maxwell–Boltzmann distribution. Since we have now an equation for the temporal evolution of f, we are in the position to compute how H changes with time, at least for our perfect gas. We shall prove that $H(t)$ is monotonically decreasing from any given initial PDF far from equilibrium and eventually reaches a stationary value for the Maxwellian. To this end, it is not necessary to make any assumptions for the collision integral.

Let us first introduce the density to H:

$$h(\vec{r},t) = \int_{-\infty}^{\infty} d^3\vec{v}\, f(\vec{v},\vec{r},t) \ln f(\vec{v},\vec{r},t), \tag{8.34}$$

which can be considered as a negative entropy density. We differentiate (8.34) with respect to time and use (8.18)

$$\begin{aligned} \frac{\partial h}{\partial t} &= \int_{-\infty}^{\infty} d^3\vec{v}\, \frac{\partial f}{\partial t}(\ln f + 1) \\ &= -\underbrace{\int_{-\infty}^{\infty} d^3\vec{v} \sum_i^3 v_i \frac{\partial f}{\partial x_i}(\ln f + 1)}_{=J_1} - \underbrace{\int_{-\infty}^{\infty} d^3\vec{v} \sum_i^3 a_i \frac{\partial f}{\partial v_i}(\ln f + 1)}_{=J_2} + \underbrace{\int_{-\infty}^{\infty} d^3\vec{v} \frac{\delta f}{\delta t}(\ln f + 1)}_{=J_3}\,. \end{aligned} \tag{8.35}$$

The first integral can be written as

$$J_1 = \int_{-\infty}^{\infty} d^3\vec{v} \sum_i^3 \frac{\partial (v_i f \ln f)}{\partial x_i} = \nabla \cdot \underbrace{\int_{-\infty}^{\infty} d^3\vec{v} \sum_i^3 \vec{v} f \ln f}_{=\vec{j}_h} = \nabla \cdot \vec{j}_h\,,$$

where j_h can be interpreted as (negative) entropy flux density. To evaluate J_2, we apply Gauss divergence theorem,

$$J_2 = \int_{-\infty}^{\infty} d^3\vec{v} \sum_i^3 \frac{\partial (a_i f \ln f)}{\partial v_i} = \lim_{|v|\to\infty} \int d^2\Omega\, |v|^2 (\hat{v}_r \cdot \vec{a})\, f \ln f = 0$$

with $\hat{v}_r$ as the radial unit vector in v-space. J_2 vanishes, because f is zero for $|v| \to \infty$.

To evaluate J_3 we write it in a bit more general form:

$$J_3 = \int_{-\infty}^{\infty} d^3\vec{v}\,\varphi(\vec{v})\frac{\delta f}{\delta t} = \int_{-\infty}^{\infty} \varphi(\vec{v}) W\,(f'f_1' - f f_1)\underbrace{d^3\vec{v}\,d^3\vec{v}_1\,d^3\vec{v}'\,d^3\vec{v}_1'}_{\equiv d^{12}v}$$

$$= \int_{-\infty}^{\infty} W\,(\varphi f' f_1' - \varphi' f' f_1')d^{12}v = \int_{-\infty}^{\infty} W f' f_1' (\varphi - \varphi')d^{12}v,$$

where $\varphi = \varphi(\vec{v})$, $\varphi' = \varphi(\vec{v}')$, $\varphi_1 = \varphi(\vec{v}_1)$, $\varphi_1' = \varphi(\vec{v}_1')$. Later we may put $\varphi(\vec{v}) = \ln f + 1$. The collision process must be the same if we interchange $\vec{v}$ with $\vec{v}_1$ and/or $\vec{v}'$ with $\vec{v}_1'$. Thus we may write as well

$$J_3 = \int_{-\infty}^{\infty} W f' f_1' (\varphi_1 - \varphi_1')d^{12}v,$$

and symmetrize

$$\boxed{\int_{-\infty}^{\infty} d^3\vec{v}\,\varphi\frac{\delta f}{\delta t} = \frac{1}{2}\int_{-\infty}^{\infty} W f_1' f'\,(\varphi + \varphi_1 - \varphi' - \varphi_1')d^{12}v}\,. \tag{8.36}$$

This is an important and useful formula. It says if $\varphi + \varphi_1$ is one of the four conserved quantities (8.20) (or a linear combination of them), one has immediately

$$\varphi + \varphi_1 = \varphi' + \varphi_1',$$

and the mean of the collision integral over this φ vanishes, whatever the PDF f looks like, and whatever the particular collision process described by W is.

Next we put $\varphi(\vec{v}) = \ln f(\vec{v}) + 1$ and obtain

$$J_3 = \frac{1}{2}\int_{-\infty}^{\infty} W f' f_1' (\ln f + \ln f_1 - \ln f' - \ln f_1')d^{12}v = -\frac{1}{2}\int_{-\infty}^{\infty} W f' f_1' \ln\frac{f' f_1'}{f f_1} d^{12}v \equiv \sigma\,. \tag{8.37}$$

Now, (8.35) can be written as

$$\frac{\partial h}{\partial t} = -\nabla\cdot\vec{j}_h + \sigma\,. \tag{8.38}$$

This has the form of a conservation equation with additional source term σ, defined in (8.37); see Section 2.1.3 and equation (2.8). After integrating over the volume, we find

$$\frac{dH}{dt} = -\frac{1}{2}\int_V d^3r \int_{-\infty}^{\infty} d^{12}v\, W f' f_1' \ln\frac{f' f_1'}{f f_1}\,. \tag{8.39}$$

The precondition for the validity of the H-Theorem is a closed system, so we assumed that the entropy flux through the surface $F(V)$ is zero:

$$\int_V d^3\vec{r}\,\nabla\cdot\vec{j}_h = \oint_{F(V)} d^2\vec{f}\cdot\vec{j}_h = 0\,.$$

Continuing with (8.39), we may interchange the velocities and symmetrize to

$$\begin{aligned}\frac{dH}{dt} &= -\frac{1}{4}\int_V d^3r \int_{-\infty}^{\infty} d^{12}v\, W\left[f'f_1'\ln\frac{f'f_1'}{f f_1} + f f_1 \ln\frac{f f_1}{f'f_1'}\right] \\ &= -\frac{1}{4}\int_V d^3r \int_{-\infty}^{\infty} d^{12}v\, W f f_1 (q-1)\ln q \end{aligned}\tag{8.40}$$

with

$$q = \frac{f'f_1'}{f f_1} \ge 0\,.$$

Now consider the integrand in (8.40). Clearly W and f are non-negative functions. But the same is true for $(q-1)\ln q$ (Figure 8.4). So we proved the H-Theorem in the same way as already Boltzmann did:

In a closed system, $H(t)$ is a monotonically decreasing function and reaches its minimum asymptotically in time. Then the gas is in equilibrium.

The minimum is at $q = 1$, thus

$$f f_1 = f'f_1'\,.\tag{8.41}$$

The PDF that fulfills (8.41) is the equilibrium distribution f^{eq}.

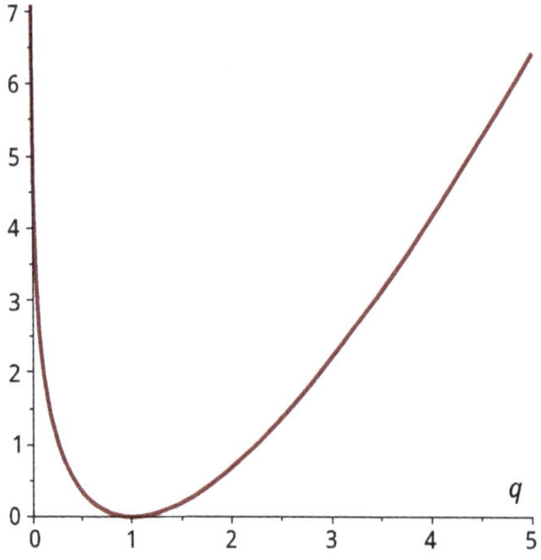

Figure 8.4: The function $(q-1)\ln q$ is non-negative and has its only minimum at $q = 1$.

8.2.5 Loschmidt's paradox

Before we continue, let us allow for a somewhat more philosophical insertion that the reader can skip without disadvantage for the remaining parts.

A serious objection against Boltzmann's reasoning was raised by Josef Loschmidt[1] in 1876, nowadays known as "Loschmidt's paradox," "irreversibility paradox" or "Umkehreinwand." Like Boltzmann, Loschmidt assumed a classical gas of particles far from equilibrium at time t_0 evolving via collisions towards equilibrium until a certain time t_1 so that $H(t_1) < H(t_0)$ according to the H-Theorem. Then at t_1 all velocities are reversed by some "demon" and all particles move backwards. If all collision processes would be reversible, as assumed by Boltzmann, the further evolution of the gas would go exactly backwards to the very same state at $t_2 = 2t_1 - t_0$, which it had already at t_0. But this state would be farther from equilibrium as that at t_1, and thus $H(t_2) > H(t_1)$, in contradiction to the H-Theorem. Loschmidt used this idea to find a back door from the road to heat death, which was prescribed by Boltzmann for the universe as a closed system.

On the other hand, if the whole world would be described on the microscopic level by classical mechanics, reversing all velocities at a certain time would correspond to "undoing what was happened" or "undoing of history." Apart from that, the state after reversing all velocities can be considered as another initial condition responsible for *an evolution out of equilibrium* and $d_t H(t) > 0$, in contradiction with the H-Theorem. But to find such a state is highly improbable compared to other states that all decrease $H(t)$ if time goes on.

As an example, consider the pool game. If we neglect friction and consider all collisions between the balls reversible, the initial condition ($t = 0$), that is, 15 balls are placed in a triangle somewhere in the middle of the table and hit by the white ball, giving them certain different velocities so that they move apart. Now, say after one second at $t = t_1$, our demon reverses all velocities. If this state is chosen as new initial condition, another second later at t_2, the 15 balls should again form the initial triangle from $t = 0$ and give all their momenta to the white ball moving backwards (undoing what was happened). Of course this is a solution of Newton's law of motion, but it depends precisely on the new initial condition at $t = t_1$, obtained after reversing all velocities. If one only changes one of the velocities (or positions) very slightly, one will end at t_2 in a completely different state, which has nothing to do with the first (triangle) initial state at $t = 0$; compare Figure 8.5.

This is exactly the central idea of chaos theory, which of course was not (or by Boltzmann perhaps only "intuitively") known in the time of Loschmidt. Arbitrarily, small perturbations on the microscopic level may lead to a completely different behavior after a certain time on the macroscopic level, if the system evolves chaotically (butterfly effect). The trajectories of the unperturbed and the perturbed system in phase space diverge

1 Johann Josef Loschmidt, Austrian physicist and chemist, 1821–1895.

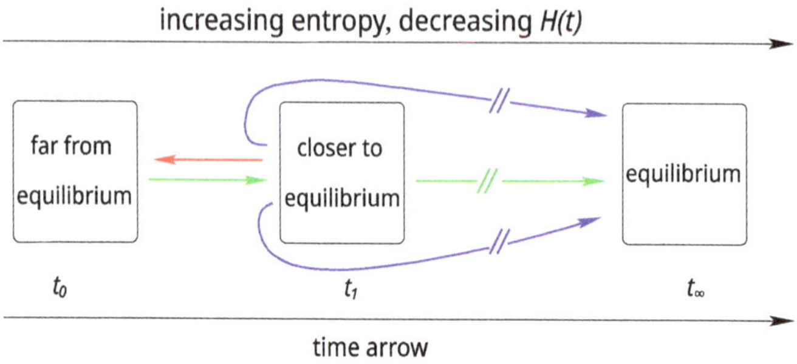

Figure 8.5: Green arrows: a closed system evolves towards equilibrium. Red arrow: if at t_1 all velocities are exactly reversed, the system evolves backwards to the initial state, time runs apparently backwards, "undoing what was happened." However, this evolution has almost zero probability. If only one of the particle velocities (or positions) is arbitrarily small perturbed, entropy indeed could decrease for a (very) short time but then increase again (blue arrows), because the path in phase space becomes completely different (assumption of molecular chaos).

on a time scale proportional to the inverse of the largest Lyapunov exponent. A positive Lyapunov exponent is a measure for chaos and can be huge for a many-particle system [5]. Chaotic evolution can take place already for a small number of interacting particles, only three at a minimum, but surely for 16, as for the billiard game. In a gas, we have some 10^{26} particles and finding an initial condition that leads away from equilibrium (increasing $H(t)$ and decreasing entropy) for a certain finite time is practically impossible.

8.2.6 Equilibrium distribution

From the condition (8.41), the Maxwell–Boltzmann distribution (7.19), already derived earlier in a completely different way, follows. Forming the logarithm of (8.41) yields

$$\underbrace{\ln f^{\mathrm{eq}} + \ln f_1^{\mathrm{eq}}}_{\text{before collision}} = \underbrace{\ln f^{\mathrm{eq}'} + \ln f_1^{\mathrm{eq}'}}_{\text{after collision}} .$$

In other words, if the system is in equilibrium, $\ln f^{\mathrm{eq}} + \ln f_1^{\mathrm{eq}}$ is a conserved quantity, called *summation invariant*. But we already know all possible conserved quantities; see (8.20). Thus the logarithm of the equilibrium PDF must be a linear combination of energy, the three momentum components and the fifth conserved quantity, namely the number of particles (or the total mass). So we may write

$$\ln f^{\mathrm{eq}}(\vec{r}, \vec{v}, t) = \alpha_m + \alpha_x v_x + \alpha_y v_y + \alpha_z v_z + \alpha_e v^2 ,$$

where $\alpha_i = \alpha_i(\vec{r}, t)$ if the gas is only in local equilibrium. This is far from looking like the Maxwellian (7.19), but changing the variables into

$$
\begin{aligned}
a_m &= \frac{3}{2}\ln\left(\frac{m}{2\pi k_B T}\right) - \frac{mu^2}{k_B T},\\
a_i &= \frac{mu_i}{k_B T}, \quad i = x, y, z,\\
a_e &= -\frac{m}{2k_B T}
\end{aligned}
$$

we arrive exactly at (7.19) or (7.20).

This is a very important result: The Maxwell–Boltzmann distribution is the one where the collision integral vanishes,

$$\frac{\delta f}{\delta t} = 0 \tag{8.42}$$

and the H-functional takes a minimum (the entropy a maximum). This distribution must then be the equilibrium PDF. If f depends also on $\vec{r}$, the equilibrium is local ((8.42) vanishes at every place $\vec{r}$); if f depends on $\vec{v}$ only, we have global equilibrium. A vanishing collision integral does of course not mean there are no collisions at all. It just says that in equilibrium depletion (8.14) and replenishment (8.15) are in balance.

In summary, the collision integral momentum

$$\int_{-\infty}^{\infty} d^3\vec{v}\, \frac{\delta f}{\delta t}\, \varphi(\vec{v}) \tag{8.43}$$

can vanish for two reasons:

(1) The gas is in (local) equilibrium, and f is a Maxwell–Boltzmann PDF. Then φ can be an arbitrary function of $\vec{v}$.

(2) The function $\varphi(\vec{v})$ is a summation invariant (mass, energy, momentum or linear combinations among them). Then f can be an arbitrary (non-equilibrium) distribution function.

The first point is a consequence of (8.41); the second one follows directly from (8.36). From (2), we may conclude with $\varphi = 1$

$$\int_{-\infty}^{\infty} d^3\vec{v}\, \frac{\delta f}{\delta t} = 0$$

for arbitrary distribution functions f.

8.2.7 Example: barometric formula

What happens if we insert the local Maxwellian (7.20) with $\vec{u} = 0$ into the Boltzmann equation (8.18)? After some computations (remember that the collision integral is zero) we see that the conditions

$$\nabla T = 0, \quad \partial_t T = 0, \quad \partial_t n = 0, \quad \nabla n = \frac{nm}{k_B T}\,\vec{a} \tag{8.44}$$

must hold. The gas is stationary, isothermal, and the density gradient must be in balance with the external forces. If we assume the gas particles in a vertical gravitation field

$$\vec{a} = -g\hat{e}_z = -\frac{1}{m}\,\nabla U$$

with the potential energy $U = mgz$, we have $n = n(z)$ and

$$\frac{dn}{dz} = -\frac{n}{k_B T}\,\frac{dU}{dz},$$

which can be integrated to the well-known *barometric formula*

$$n(z) = n(0)\exp\left(-\frac{U}{k_B T}\right) = \exp\left(-\frac{mg}{k_B T}z\right). \tag{8.45}$$

The density is exponentially decreasing with increasing height in an isothermal atmosphere, Figure 8.6. Using the perfect gas equation, the same behavior is found for the pressure.

Figure 8.6: Particles of an isothermal gas in local equilibrium under a constant vertical gravitational acceleration.

However, the atmosphere is far from being isothermal; the temperature decreases with increasing height by roughly 0.01 K/m in the lower atmosphere. To describe such a situation, the local Maxwell–Boltzmann distribution is obviously not sufficient, and the collision integral must also play an important role for thermal conduction and thermalization. We shall see in the next chapter how the local Maxwell–Boltzmann PDF can be extended systematically to account for more involved solutions.

With (8.44), it is interesting to see that a local Maxwellian with (arbitrarily) varying temperature, density and mean velocity is in general not a solution of the stationary Boltzmann equation.

Taking (8.45), the complete local Maxwellian (7.20) reads

$$f(z,v) = n_0 \left(\frac{m}{2\pi k_B T}\right)^{1/2} \mathrm{e}^{-(\frac{1}{2}mv^2+U)/k_BT} = n_0 \left(\frac{m}{2\pi k_B T}\right)^{1/2} \mathrm{e}^{-E/k_BT},$$

which is just the canonical Boltzmann distribution [1].

8.2.8 Mean free time again

In Section 7.4.1, we estimated the mean free time between two particle collisions. Now we can use the depletion part of the collision integral in the form of (8.28) to find a better estimate. Integrating (8.28) over $\vec{v}$ gives the number of collisions ΔN_c per time Δt as

$$\frac{\Delta N_c}{\Delta t} n = \int d^3\vec{v}\, d^3\vec{v}_1 d^2\Omega' \; |\vec{v}-\vec{v}_1| \; S(\vartheta')f(\vec{v})f(\vec{v}_1)\,. \tag{8.46}$$

Then, the reciprocal

$$\tau = \left(\frac{\Delta N_c}{\Delta t}\right)^{-1}$$

is the average time between two collisions, and therefore equal to the mean free time. To compute (8.46), we can assume a Maxwellian for f, leading to

$$\frac{\Delta N_c}{\Delta t} = \frac{n\sigma m^3}{(2\pi k_B T)^3} \int d^3\vec{v}\, d^3\vec{v}_1 \, |\vec{v}-\vec{v}_1| \; \exp\left(-\frac{m(v^2+v_1^2)}{2k_BT}\right), \tag{8.47}$$

where we have already integrated over the solid angle, and

$$\sigma = \int d^2\Omega\, S(\vartheta)$$

is the total scattering angle. In the case of hard spheres, one finds $\sigma = 4\pi R^2$. The integrals in (8.47) can be computed easily by transforming to relative and mass centered coordinates (8.21) according to

$$\frac{\Delta N_c}{\Delta t} = \frac{4\pi n\sigma m^3}{(2\pi k_B T)^3} \int d^3\vec{v}_c \, \exp\left(-\frac{mv_c^2}{k_BT}\right) \int_0^\infty dv_r v_r^3 \, \exp\left(-\frac{mv_r^2}{4k_BT}\right) = 4\sqrt{\frac{k_BT}{\pi m}}\, n\sigma\,,$$

leading finally to

$$\tau = \frac{1}{4n\sigma}\sqrt{\frac{\pi m}{k_B T}} .$$

Comparing this with our former result (7.54), the new value is larger by a factor $\sqrt{3\pi/8} \approx 1.085$.

9 Transport equations

In this chapter, we wish to give a statistical foundation of the phenomenologically derived transport equations for mass, momentum, and energy in Chapter 3. The derivation is based on the Boltzmann equation (8.18), and thus founded on first principles of classical (Newtonian) particle mechanics.

9.1 Maxwell's equation of change

9.1.1 General form

Density, momentum, and kinetic energy can be expressed as moments of the PDF; see Section 7.2.1. We take the general form (7.4)

$$\langle\varphi(\vec{v})\rangle = \int_{-\infty}^{\infty} d^3\vec{v}\;\varphi(\vec{v})\,f(\vec{r},\vec{v},t) \tag{9.1}$$

with a scalar function φ depending on $\vec{v}$ only. Multiplying (8.18) with φ and integrating over $\vec{v}$ yields (all integrals from $-\infty$ to ∞)

$$\underbrace{\int d^3\vec{v}\,\varphi\frac{\partial f}{\partial t}}_{=J_1} + \underbrace{\int d^3\vec{v}\,\varphi\sum_{i=1}^{3} a_i\frac{\partial f}{\partial v_i}}_{=J_2} + \underbrace{\int d^3\vec{v}\,\varphi\sum_{i=1}^{3} v_i\frac{\partial f}{\partial r_i}}_{=J_3} = \int d^{12}v\,\varphi\,W\,(f'f_1' - f f_1)\,. \tag{9.2}$$

The integrals on the l. h. s. are converted as

$$J_1 = \frac{\partial}{\partial t}\int d^3\vec{v}\,\varphi f = \frac{\partial\langle\varphi\rangle}{\partial t}\,,$$

$$J_2 = -\int d^3\vec{v}\sum_{i=1}^{3} a_i f\frac{\partial\varphi}{\partial v_i} = -\sum_{i=1}^{3} a_i\left\langle\frac{\partial\varphi}{\partial v_i}\right\rangle,$$

$$J_3 = \sum_{i=1}^{3}\frac{\partial}{\partial r_i}\int d^3\vec{v}\,\varphi\,v_i\,f = \sum_{i=1}^{3}\frac{\partial}{\partial r_i}\langle v_i\varphi\rangle = \nabla\cdot\langle\vec{v}\varphi\rangle\,,$$

where in J_2 integration by parts has been applied, and a_i is supposed to be independent of $\vec{v}$. Using (8.36) on the r. h. s., equation (9.2) turns into *Maxwell's equation of change*

$$\frac{\partial\langle\varphi\rangle}{\partial t} - \sum_{i=1}^{3} a_i\left\langle\frac{\partial\varphi}{\partial v_i}\right\rangle + \nabla\cdot\langle\vec{v}\varphi\rangle = \frac{1}{2}\int_{-\infty}^{\infty} W f_1' f'\,(\varphi+\varphi_1-\varphi'-\varphi_1')\,d^{12}v\,. \tag{9.3}$$

From (9.3), the spatio-temporal development of any function average $\langle\varphi\rangle$ can be computed. Note that (9.3) is a (3 + 1)-dimensional partial differential equation (space and

https://doi.org/10.1515/9783110794137-011

time); it already has the same dimensionality than the transport equations we are looking for.

Maxwell's equation of change can also be formulated with random velocities $\vec{w}$ and with $f(\vec{r}, \vec{w}, t)$. Then one takes in (9.2) the Boltzmann equation in the form of (8.19) and obtains an equation for $\langle\varphi(\vec{w})\rangle$. The conversions are essentially the same. We present here only the result for later purpose:

$$\begin{aligned}\frac{\partial\langle\varphi\rangle}{\partial t} &- \sum_{i=1}^{3} a_i\left\langle\frac{\partial\varphi}{\partial w_i}\right\rangle + \sum_{i=1}^{3}\frac{\partial}{\partial r_i}(\langle w_i\varphi\rangle + u_i\langle\varphi\rangle)\\ &+ \sum_{i=1}^{3}\left\langle\frac{\partial\varphi}{\partial w_i}\right\rangle\left(\frac{\partial u_i}{\partial t} + \sum_{j=1}^{3} u_j\frac{\partial u_i}{\partial r_j}\right) + \sum_{i,j=1}^{3}\left\langle w_j\frac{\partial\varphi}{\partial w_i}\right\rangle\frac{\partial u_i}{\partial r_j}\\ &= \frac{1}{2}\int_{-\infty}^{\infty} W f_1' f' \left(\varphi + \varphi_1 - \varphi' - \varphi_1'\right) d^{12}w\,. \end{aligned} \tag{9.4}$$

9.1.2 Summation invariants

The evaluation of (9.3) or (9.4) is particularly simple if φ is one of the five summation invariants, because then the collision integral is zero. We begin with the following:

(i) **Mass:**

$$\varphi = m, \quad \langle\varphi\rangle = mn, \quad \left\langle\frac{\partial\varphi}{\partial v_i}\right\rangle = 0\,,$$

and (9.3) turns into

$$m\frac{\partial n}{\partial t} + m\sum_{i=1}^{3}\frac{\partial}{\partial r_i}(nu_i) = m\frac{\partial n}{\partial t} + m\nabla\cdot(n\vec{u}) = 0\,. \tag{9.5}$$

With $mn = \rho$ this is the continuity equation (2.11) or (3.1). The PDF is not needed.

(ii) **Momentum:**

$$\varphi = v_i, \quad \langle\varphi\rangle = nu_i, \quad \left\langle\frac{\partial\varphi}{\partial v_i}\right\rangle = n\,,$$

and from (9.3)

$$\frac{\partial(nu_i)}{\partial t} - a_i n + \sum_{j=1}^{3}\frac{\partial}{\partial r_j}\langle v_i v_j\rangle = 0\,. \tag{9.6}$$

For closure, we must represent $\langle v_i v_j\rangle$ as functions of the macroscopic variables $\vec{u}$, n, and p. Taking the random or thermal velocities $\vec{w}(\vec{r}, t) = \vec{v} - \vec{u}(\vec{r}, t)$ defined in (7.7), we write

$$\langle v_i v_j \rangle = \langle (w_i + u_i)(w_j + u_j) \rangle = \langle w_i w_j \rangle + n u_i u_j \,. \tag{9.7}$$

To compute $\langle w_i w_j \rangle$, however, the PDF is needed (see below). Here we can only say that the diagonal moments $\langle w_i w_i \rangle$ are related to the pressure from equation (7.10), and that due to isotropy they must be all equal:

$$\langle w_x w_x \rangle = \langle w_y w_y \rangle = \langle w_z w_z \rangle = \frac{1}{3}\langle w^2 \rangle = \frac{p}{m} \,. \tag{9.8}$$

To extract the pressure, we define the second moment tensor $\underline{\tau}$ according to

$$\tau_{ij} = -m \langle w_i w_j \rangle + \delta_{ij}\, p \,. \tag{9.9}$$

Inserting (9.7) with (9.9) into (9.6) yields after some further manipulation the transport equation

$$\rho \left[\frac{\partial \vec{u}}{\partial t} + (\vec{u} \cdot \nabla)\vec{u} \right] = \nabla \cdot \underline{\tau} + \vec{f} - \nabla p \tag{9.10}$$

with density $\rho = nm$ and $\vec{f} = \rho\, \vec{a}$ being the volume force. Note the similarity of (9.10) with the momentum balance equation (2.23b). From there it is clear that $\underline{\tau}$ is related to the viscous stress tensor. Using the Maxwell–Boltzmann PDF, $\underline{\tau} = 0$ and the Euler equations for an inviscid fluid are recovered.

(iii) **Internal energy (or pressure or temperature):**

$$\varphi = w^2, \quad \langle \varphi \rangle = \frac{3p}{m}, \quad \left\langle \frac{\partial \varphi}{\partial w_i} \right\rangle = 2\,\langle w_i \rangle = 0 \,.$$

Now φ is a function of $\vec{w}$, and we must take equation (9.4) instead of (9.3). Using (9.9), (9.4) can be cast into a conservation equation for $\langle w^2 \rangle$ with source terms on the r. h. s.:

$$\frac{\partial \langle w^2 \rangle}{\partial t} + \nabla \cdot (\vec{u}\,\langle w^2 \rangle) = -\frac{2}{m}\,\nabla \cdot \vec{q} - \frac{2}{3}\langle w^2 \rangle\, \nabla \cdot \vec{u} + 2\,\underline{\tau} : (\nabla \vec{u}) \,, \tag{9.11}$$

where

$$q_i = \frac{m}{2}\,\langle w_i w^2 \rangle \tag{9.12}$$

will play the role of the heat flux (later we shall compute it, but also here the PDF is needed). For any symmetric (even) PDF with $f(\vec{w}) = f(-\vec{w})$, especially for the Maxwell–Boltzmann PDF, one finds $\vec{q} = 0$.

From equation (9.8), we may substitute $\langle w^2 \rangle = 3p/m$ and obtain the pressure equation for a monatomic gas under adiabatic conditions (symmetric PDF, $\vec{q} = 0$, $\underline{\tau} = 0$):

$$\frac{\partial p}{\partial t} + \nabla \cdot (\vec{u}\, p) + \frac{2}{3}\, p\, \nabla \cdot \vec{u} = 0 \,. \tag{9.13}$$

But we can also use the relation (7.11), $\langle w^2 \rangle = 3k_B Tn/m$, and find from (9.11) the heat equation according to

$$\frac{3k_B n}{2}\left[\frac{\partial T}{\partial t} + \vec{u}\cdot\nabla T\right] = -\nabla\cdot\vec{q} - p\,\nabla\cdot\vec{u} + m\,\underline{\tau} : (\nabla\vec{u})\,. \tag{9.14}$$

If we compare this equation with the heat equation (3.8), we see immediately that $\vec{q}$ is related to the heat flux.

We repeat the very important finding from this last paragraph:

In local equilibrium the PDF is a local Maxwellian, heat flux and the shear stress tensor both vanish.

9.2 Symmetric PDF, Euler equations

The equations (9.10), (9.11) contain higher moments and are not closed. Using Maxwell's equation of change, one may write down PDEs for these higher moments, but they would include even higher moments, and so on. So one must try to systematically close this hierarchy. But before we do this, we consider the most simple case of a symmetric (even) PDF, where $\vec{q} = \underline{\tau} = 0$.

9.2.1 Local equilibrium

Here we shall summarize the lowest approximation that is obtained assuming for instance for f a local Maxwellian. Then, $\underline{\tau}$ and $\vec{q}$ vanish, and the five PDEs for the summation invariants (9.5), (9.10), (9.11) are closed. We repeat them here, writing density ρ instead of mn:

$$\frac{\partial\rho}{\partial t} + \nabla\cdot(\rho\vec{u}) = 0, \tag{9.15a}$$

$$\rho\left[\frac{\partial\vec{u}}{\partial t} + (\vec{u}\cdot\nabla)\vec{u}\right] = \vec{f} - \nabla p, \tag{9.15b}$$

$$\frac{\partial p}{\partial t} + \nabla\cdot(\vec{u}p) = -\frac{2}{3}\,p\,\nabla\cdot\vec{u}\,. \tag{9.15c}$$

These are the Euler equations, five equations for the five macroscopic variables

$$\rho(\vec{r},t), \quad \vec{u}(\vec{r},t), \quad p(\vec{r},t)\,.$$

They can be completed by a thermal equation of state $p = p(\rho, T)$, for instance the perfect gas equation:

$$p(\rho, T) = \frac{\rho k_B T}{m}, \tag{9.16}$$

from where the temperature follows. Alternatively, one may use the temperature equation (9.14)

$$C\rho \left[\frac{\partial T}{\partial t} + \vec{u} \cdot \nabla T \right] = -p \, \nabla \cdot \vec{u} \tag{9.17}$$

with $C = 3k_B/2m$ instead of (9.15c), and then compute the pressure from (9.16).

Obviously, the knowledge of the collision physics is not needed to derive this set. This is expressed in the fact that diffusion does not occur; neither that of momentum ($\underline{\tau} = 0$) nor that of temperature (or pressure, $\vec{q} = 0$). To cover diffusion, the assumption of local equilibrium must be extended to PDFs, which deviate from a Maxwellian, as we shall see in Section 9.3 below.

9.2.2 Isothermal and incompressible conditions

The Euler equations are also a good approximation for fluids with low viscosity, such as water. If the fluid has in addition a high thermal conductivity, the isothermal approximation

$$T = \text{const.} \tag{9.18}$$

is often used. For an incompressible flow, $\nabla \cdot \vec{u} = 0$ and (9.18) is an exact solution of (9.17). Though temperature is not included explicitly in (9.15), the system of five equations plus state equation is over-determined with (9.18). Then one uses only (9.15a) and (9.15b) and determines the pressure from (9.16) (or from some similar EOS for a liquid) together with (9.18).

On the other hand liquids are to a great extent incompressible,

$$\rho = \rho_0 = \text{const.}$$

Then the only equations left are

$$\nabla \cdot \vec{u} = 0, \tag{9.19a}$$

$$\rho_0 \left[\frac{\partial \vec{u}}{\partial t} + (\vec{u} \cdot \nabla)\vec{u} \right] = \vec{f} - \nabla p, \tag{9.19b}$$

and a state equation is no longer needed. There exist several methods to eliminate or to determine the pressure implicitly to fulfill (9.19a); see textbooks on fluid dynamics [8, 9], but also the remarks of Section 3.1.2 and equation (3.7).

9.2.3 Application: sound waves

Sound waves in gases or liquids are dilatational pressure waves of normally small amplitude. One can linearize

$$p = p_0 + \tilde{p}, \quad \rho = \rho_0 + \tilde{\rho}$$

and obtains from (9.15) ($\vec{f} = 0$):

$$\frac{1}{\rho_0}\frac{\partial \tilde{\rho}}{\partial t} = -\nabla \cdot \vec{u}, \tag{9.20a}$$

$$\rho_0 \frac{\partial \vec{u}}{\partial t} = -\nabla \tilde{p}, \tag{9.20b}$$

$$\frac{1}{p_0}\frac{\partial \tilde{p}}{\partial t} = -\frac{5}{3}\nabla \cdot \vec{u}\,. \tag{9.20c}$$

It is clear that sound waves can only exist if $\nabla \cdot \vec{u} \neq 0$, i. e., in a compressible medium. Differentiating (9.20c) by time and inserting (9.20b) yields a wave equation for the pressure

$$\frac{\partial^2 \tilde{p}}{\partial t^2} - \frac{5}{3}\frac{p_0}{\rho_0}\nabla^2 \tilde{p} = 0\,. \tag{9.21}$$

From there one can read off the speed of sound

$$c_s = \left(\frac{5}{3}\frac{p_0}{\rho_0}\right)^{1/2}. \tag{9.22}$$

Note that (9.15c) is derived for a monatomic gas. The general formula for c_s found by Laplace in 1816 reads

$$c_s = \left(\gamma \frac{p_0}{\rho_0}\right)^{1/2}$$

with the adiabatic index γ, which is 5/3 for a monatomic and 7/5 for a diatomic gas, such as air.

Isaac Newton was the first who tried to compute the speed of sound in air already in 1710. However, he assumed the air being isothermal and used the perfect gas equation to express

$$\tilde{p}(\tilde{\rho}) = p_0 \frac{\tilde{\rho}}{\rho_0}\,.$$

Eliminating $\vec{u}$ from (9.20a) and (9.20b) yields

$$\frac{\partial^2 \tilde{p}}{\partial t^2} - \frac{p_0}{\rho_0} \nabla^2 \tilde{p} = 0 \tag{9.23}$$

and

$$c_s = \left(\frac{p_0}{\rho_0}\right)^{1/2},$$

a discrepancy by a factor of $\sqrt{5/7} \approx 0.85$.

9.3 Non-equilibrium distribution function

To include diffusion processes in the transport equations, the Maxwell–Boltzmann PDF has to be systematically extended to a non-equilibrium and non-symmetric PDF, containing also odd terms in the components of $\vec{v}$. The idea is to expand the non-equilibrium PDF with respect to the velocities. We present the derivation in random variables. Let

$$f^M(\vec{r}, \vec{w}, t) = n(\vec{r}, t)\left(\frac{m}{2\pi k_B T(\vec{r}, t)}\right)^{3/2} \exp\left[-\frac{m w(\vec{r}, t)^2}{2k_B T(\vec{r}, t)}\right]$$

be a local Maxwellian. Then the non-equilibrium PDF is written as a series:

$$f(\vec{r}, \vec{w}, t) = f^M(\vec{r}, \vec{w}, t)\left[1 + \sum_{i=1}^{3} \alpha_i(\vec{r}, t) w_i + \sum_{i,j=1}^{3} \beta_{ij}(\vec{r}, t) w_i w_j + \sum_{i,j,\ell=1}^{3} \gamma_{ij\ell}(\vec{r}, t) w_i w_j w_\ell + \cdots\right]. \tag{9.24}$$

This is the Chapman–Enskog expansion. If the system is not far from local equilibrium, the series (9.24) should quickly converge. Instead of (9.24), we shall write

$$f(\vec{r}, \vec{w}, t) = f^M(\vec{r}, \vec{w}, t) + \varepsilon f^{(1)}(\vec{r}, \vec{w}, t) + \varepsilon^2 f^{(2)}(\vec{r}, \vec{w}, t) + \cdots. \tag{9.25}$$

with $\varepsilon \ll 1$.

9.3.1 Stress tensor

We start with the momentum conservation formulated as (9.10). For closure, we must express $\tau_{ij} = -\langle w_i w_j \rangle + \delta_{ij} p/m$ as function of $\vec{u}$, n and p. To this end, we evaluate Maxwell's equation of change (9.4) with

$$\varphi(\vec{w}) = w_k w_\ell$$

and with the PDF written as (9.25). We restrict the treatment to the lowest order and neglect all terms that are of $O(\varepsilon)$. For (9.4), we need to compute the following moments:

$$\langle w_k w_\ell \rangle = \delta_{k\ell} \frac{p}{m} + O(\varepsilon),$$
$$\left\langle \frac{\partial w_k w_\ell}{\partial w_i} \right\rangle = \delta_{ki} \langle w_\ell \rangle + \delta_{\ell i} \langle w_k \rangle = O(\varepsilon),$$
$$\langle w_i w_k w_\ell \rangle = O(\varepsilon),$$
$$\left\langle w_j \frac{\partial w_k w_\ell}{\partial w_i} \right\rangle = \delta_{ki} \langle w_j w_\ell \rangle + \delta_{\ell i} \langle w_j w_k \rangle = (\delta_{j\ell} \delta_{ki} + \delta_{jk} \delta_{\ell i}) \frac{p}{m} + O(\varepsilon) .$$

Inserting this into (9.4) leads to

$$\delta_{ij} \left[\frac{\partial p}{\partial t} + \nabla \cdot (p \vec{u}) \right] + p \sum_{\ell,k=1}^{3} (\delta_{ik} \delta_{\ell j} + \delta_{jk} \delta_{\ell i}) \frac{\partial u_\ell}{\partial r_k} = m \, C^s_{kl}, \tag{9.26}$$

where

$$C^s_{k\ell} = \int_{-\infty}^{\infty} d^3 \vec{w} \, \frac{\delta f}{\delta t} \, w_k w_\ell \tag{9.27}$$

is a bilinear functional of f^M and $f^{(1)}$. On the l. h. s. of (9.26), we have neglected all terms of $O(\varepsilon)$. This of course is only consistent if we assume that the collision integral (9.27) is of order one, as we shall discuss below.

The terms in the square brackets of (9.26) can be replaced up to the same order using equation (9.13). Finally, we can evaluate the double sum and find

$$-\frac{2}{3} \delta_{k\ell} \, p \nabla \cdot \vec{u} + p \left(\frac{\partial u_k}{\partial r_\ell} + \frac{\partial u_\ell}{\partial r_k} \right) = m \, C^s_{k\ell} \, . \tag{9.28}$$

This is the first place where the collision integral is explicitly needed.

9.3.2 Relaxation time approximation (BGK approximation)

To continue, the collision integral (9.27) has to be evaluated. A very convenient and straight-forward method to do this without the need of considering further details of the collision process was proposed be Bhatnagar, Gross, and Krook in 1954, now called BGK approximation.

The H-theorem tells us that in a closed system every initial PDF must go monotonically towards a (local) Maxwell–Boltzmann PDF. The distance in function space

$$\bar{d}(t) = \int_V d^3\vec{r} \int_{-\infty}^{\infty} d^3\vec{v}\, |f(\vec{r},\vec{v},t) - f^M(\vec{r},\vec{v},t)|$$

is a monotonically decreasing function of time and reaches zero in equilibrium for any arbitrary distribution function f. If f does not depend on space and there are no external forces a_i, the Boltzmann equation (8.18) takes the form

$$\frac{\partial f}{\partial t} = \frac{\delta f}{\delta t}, \tag{9.29}$$

and the only way to change f is due to collisions. Assuming

$$\boxed{\frac{\delta f}{\delta t} = -\frac{f - f^M}{\tau_0}} \tag{9.30}$$

defines the *collision integral in the relaxation time approximation* or in the *BGK approximation*. The relaxation time τ_0 enters here as (the only) free parameter describing the physics of collisions; we shall see later that it can be directly related to the free mean time introduced in Section 7.4.1.

The BGK approximation (9.30) inserted into (9.29) leads to

$$\frac{\partial f}{\partial t} = -\frac{f - f^M}{\tau_0} \tag{9.31}$$

and has the solution

$$f(\vec{v},t) = f^M(\vec{v}) + [f(\vec{v},t=0) - f^M(\vec{v})]\, e^{-t/\tau_0} .$$

This corresponds to an exponential relaxation of f towards f^M within the relaxation time τ_0. The collisions will render an arbitrarily distributed system to a Maxwellian. For $\tau_0 \to 0$, f turns immediately to f^M, and the gas is always in (local) equilibrium. Thus it is self-evident to assume

$$\tau_0 = O(\varepsilon) \tag{9.32}$$

with the same ε from the expansion (9.25). Inserting (9.25) into (9.30) yields

$$\frac{\delta f}{\delta t} = -\frac{\varepsilon}{\tau_0} f^{(1)} \tag{9.33}$$

and justifies with (9.32) the assumption $C^s_{k\ell}$ of order one, as already anticipated in (9.26). With (9.33), we can easily compute (9.27) as

$$\begin{aligned} C^s_{k\ell} &= -\frac{\varepsilon}{\tau_0}\int\limits_{-\infty}^{\infty} d^3\vec{w} f^{(1)}\, w_k w_\ell \\ &= -\frac{1}{\tau_0}\int\limits_{-\infty}^{\infty} d^3\vec{w} f\, w_k w_\ell - \frac{1}{\tau_0}\int\limits_{-\infty}^{\infty} d^3\vec{w} f^M\, w_k w_\ell + O(\varepsilon) \\ &= -\frac{1}{\tau_0}\left[\langle w_k w_\ell\rangle - \delta_{k\ell}\,\frac{p}{m}\right] + O(\varepsilon)\,. \end{aligned} \tag{9.34}$$

Comparing with (9.9) yields in lowest order

$$C^s_{k\ell} = \frac{1}{\tau_0}\frac{\tau_{ij}}{m}\,,$$

and with (9.28)

$$\tau_{ij} = p\,\tau_0\left[\left(\frac{\partial u_i}{\partial r_j} + \frac{\partial u_j}{\partial r_i}\right) - \frac{2}{3}\,\delta_{ij}\,\nabla\cdot\vec{u}\right]. \tag{9.35}$$

If we compare this with our phenomenological expression (2.70), we recognize exactly the same form if we identify

$$\eta = \tau_0\, p, \quad \zeta = 0\,. \tag{9.36}$$

Thus we have found a linear relation between viscosity and pressure, where the relaxation time plays the role of the proportionality constant.

It is well known that the 2nd (or volume) viscosity for a monatomic gas is zero.

9.3.3 Heat flux

The heat flux vector introduced in (9.12) and needed for the heat equation (9.14) can be computed in a similar way. Putting

$$\varphi(\vec{w}) = w_k w^2\,,$$

we need the moments that we explicitly present here again only in order one as follows:

$$\begin{aligned} \langle w_k w^2\rangle &= O(\varepsilon), \\ \left\langle \frac{\partial w_k w^2}{\partial w_i}\right\rangle &= \delta_{ki}\langle w^2\rangle + 2\langle w_i w_k\rangle = 5\delta_{ki}\frac{p}{m} + O(\varepsilon), \\ \langle w_i w_k w^2\rangle &= \delta_{ik}\langle w_i^2 w^2\rangle + O(\varepsilon) = 5\delta_{ki}\frac{p^2}{m^2 n} + O(\varepsilon), \\ \left\langle w_j \frac{\partial w_k w^2}{\partial w_i}\right\rangle &= O(\varepsilon)\,. \end{aligned}$$

Inserting this into (9.4), Maxwell's equation of change reads

$$\frac{5}{m^2}\frac{\partial}{\partial r_k}\left(\frac{p^2}{n}\right) - \frac{5p}{m}\frac{1}{\rho}\frac{\partial p}{\partial r_k} = C_k^h, \tag{9.37}$$

with

$$C_k^h = \int_{-\infty}^{\infty} d^3\vec{w}\,\frac{\delta f}{\delta t}\,w_k w^2 \tag{9.38}$$

as a bilinear functional of f and $f^{(1)}$ of order one. For the second expression on the l. h. s. of (9.37), we already used equation (9.10), where $\underline{\tau}$ must not be considered since it is in the next order. With $\rho = mn$, equation (9.37) is written as

$$\frac{5}{m^2}\left(\frac{\partial}{\partial r_k}\left(\frac{p^2}{n}\right) - \frac{p}{n}\frac{\partial p}{\partial r_k}\right) = C_k^h$$

or, using the state equation $p = nk_BT$:

$$\frac{5k_Bp}{m^2}\frac{\partial T}{\partial r_k} = C_k^h\,. \tag{9.39}$$

Again, the collision integral (9.38) can be evaluated as in (9.34) using BGK approximation:

$$C_k^h = -\frac{\varepsilon}{\tau_0}\int_{-\infty}^{\infty} d^3\vec{w}f^{(1)}\,w_k w^2 = -\frac{1}{\tau_0}\int_{-\infty}^{\infty} d^3\vec{w}f\,w_k w^2 = -\frac{1}{\tau_0}\,\langle w_k w^2\rangle\,. \tag{9.40}$$

Inserting (9.12) yields

$$C_k^h = -\frac{2q_k}{\tau_0 m}\,,$$

and with (9.39)

$$q_k = -\frac{5}{2}\frac{k_B}{m}\,p\,\tau_0\,\frac{\partial T}{\partial r_k}\,. \tag{9.41}$$

Comparing with (2.68), one obtains the thermal conductivity

$$\lambda = \frac{5}{2}\frac{k_B}{m}\,p\,\tau_0\,. \tag{9.42}$$

This means we have almost reached our goal and computed the transport coefficients from first principles, namely from the conservation equations of classical mechanics.

9.3.4 Relaxation time and mean free time

So far, we have successfully boiled down the computation of the transport coefficients to expressions containing only one free parameter, namely the relaxation time τ_0. Remember that τ_0 is the typical time that is needed for any initial PDF to reach equilibrium due to collisions. Thus it is appropriate to take τ_0 at least of the same order than the mean free time encountered in Section 7.4.1

$$\tau_0 = \alpha\,\tau = \frac{\alpha}{\sqrt{2}n\sigma|\bar{v}|}, \tag{9.43}$$

where we have introduced a numerical factor α of order one. If we take for the mean particle speed

$$|\bar{v}| = \sqrt{\frac{8k_BT}{\pi m}},$$

we find for the viscosity (with $p = nk_BT$)

$$\eta = \frac{\alpha}{4\sigma}\sqrt{\pi m k_B T} \tag{9.44}$$

and for the thermal conductivity

$$\lambda = \frac{5\alpha}{8}\frac{k_B}{\sigma}\sqrt{\frac{\pi k_B T}{m}}. \tag{9.45}$$

Comparing this last expression with our estimate (7.66) from Section 7.4.3, we find almost exactly the same up to a factor

$$\alpha = \frac{16}{5\pi} \approx 1.02\,.$$

It is interesting to see that both coefficients η and λ depend only on material properties and on $\sqrt{T}$, but not on density. This, of course, and the whole derivation, is only true for a perfect (monatomic) gas.

9.3.5 Experimental values

Note that the ratio

$$\frac{\eta}{\lambda} = \frac{2}{5}\frac{m}{k_B} = \frac{2}{5}\frac{M}{R} \tag{9.46}$$

depends only on the particle mass m or molar mass M (R = gas constant). Thus we can compare our result with measured values of monatomic gases, such as the noble gases argon and helium.

Values found from the literature are

$$\text{helium:}\quad \lambda = 0.14\,\frac{\text{W}}{\text{mK}}, \qquad \eta = 19\cdot 10^{-6}\,\frac{\text{kg}}{\text{ms}},$$

and

$$\text{argon:}\quad \lambda = 0.016\,\frac{\text{W}}{\text{mK}}, \qquad \eta = 21\cdot 10^{-6}\,\frac{\text{kg}}{\text{ms}}.$$

This gives the ratios

$$\text{helium:}\quad \eta/\lambda = 1.36\cdot 10^{-4}\,\frac{\text{Ks}^2}{\text{m}^2}; \quad \text{argon:}\quad \eta/\lambda = 1.31\cdot 10^{-3}\,\frac{\text{Ks}^2}{\text{m}^2}.$$

Computing (9.46) yields

$$\text{helium:}\quad \eta/\lambda = 1.9\cdot 10^{-4}\,\frac{\text{Ks}^2}{\text{m}^2}; \quad \text{argon:}\quad \eta/\lambda = 1.9\cdot 10^{-3}\,\frac{\text{Ks}^2}{\text{m}^2}.$$

The experimental ratios are smaller by about a factor 0.7. Keeping in mind all the approximations and assumptions done to arrive at (9.46), this seems to be a quite satisfying agreement.

An important dimensionless characteristic in non-isothermal fluid dynamics is the Prandtl number, defined as the ratio between thermal and viscous time scales,

$$P_r = \frac{\nu}{\kappa};$$

see Section 6.2. Taking $\nu = \eta/\rho$ and $\kappa = \lambda/c_p\rho$, we can write

$$P_r = \frac{\eta}{\lambda}\, c_p$$

with the specific heat per molar mass c_p. With $c_p = 5/2\cdot R/M$ and using (9.46), we find the universal value for a monatomic gas

$$P_r = 1,$$

in reasonable agreement with the literature.

9.4 Numerical solutions of the Boltzmann equation

Finally, we demonstrate with two examples how the BGK approximation can be applied to find solutions of the Boltzmann equation far from thermal equilibrium. To keep the numerical effort manageable, we shall restrict the treatment to one spatial dimension.

9.4.1 Non-dimensional Boltzmann equation

The whole computation is performed best in non-dimensional variables. To this end, we use the following scaling, where variables bearing a tilde have a physical dimension:

$$x = \frac{\tilde{x}}{\ell}, \quad t = \frac{\tilde{t}}{\tau}, \quad T = \frac{\tilde{T}}{T_0}, \quad v = \tilde{v}\,\frac{\tau}{\ell}, \quad n = \frac{\tilde{n}\ell}{N} .$$

Here, N denotes the total number of particles, T_0 a reference temperature, ℓ the length of the system, and

$$\tau = \ell\sqrt{\frac{m}{k_B T_0}} = \ell\sqrt{\frac{M}{RT_0}}$$

the (thermal) time scale. The one-dimensional Maxwell–Boltzmann distribution takes the form

$$f^M(x, v, t) = \frac{n(x,t)}{\sqrt{2\pi T(x,t)}} \exp\left[-\frac{(v - u(x,t))^2}{2T(x,t)}\right] . \tag{9.47}$$

For a numerical realization, the interval of the velocity must be limited:

$$-v_m \le v \le v_m ,$$

from (9.47), it is justified to taking $v_m \approx 10$.

9.4.2 Algorithm

We only consider the case without external forcing, $a = 0$. The non-dimensional 1D Boltzmann equation with BGK collision term reads

$$\frac{\partial f}{\partial t} + v\,\frac{\partial f}{\partial x} = \frac{f^M - f}{\tau_0} . \tag{9.48}$$

The Maxwellian f^M has to be constructed in such a way that the moments of the collision term vanish for each of the five summation invariants for all times. It follows that

$$\langle 1\rangle^M = \langle 1\rangle, \quad \langle u\rangle^M = \langle u\rangle, \quad \left\langle w^2\right\rangle^M = \left\langle w^2\right\rangle \tag{9.49}$$

with $\langle\ldots\rangle^M \equiv \int d^3\vec{v} \ldots f^M$. The general concept solving the Boltzmann equation for space dependent PDFs is the following:

(i) Take a certain initial PDF, $f(t_0) = f_0$. This can be for instance a local Maxwellian, which fulfills certain initial conditions for temperature, density, and velocity.

(ii) Determine instant density, velocity, temperature from f by

$$n(x) = \langle 1 \rangle, \quad u(x) = \frac{1}{n(x)} \langle v \rangle, \quad T(x) = \frac{1}{n(x)} \left\langle (v-u)^2 \right\rangle .$$

(iii) Compute the local Maxwellian

$$f^M(x,v) = \frac{n(x)}{\sqrt{2\pi T(x)}} \exp\left[-\frac{(v-u(x))^2}{2T(x)} \right],$$

which fulfills the relations (9.49) at time t.

(iv) Integrate numerically (see below) the Boltzmann equation (9.48) for a small time step Δt with the collision term according to (9.30); this yields

$$f(t_0 + \Delta t) .$$

(v) Go to (ii) and compute the new moments with $f(t_0 + \Delta t)$.

The Boltzmann equation (9.48) can be solved numerically by a finite difference method on a $I \times J$ grid with the mesh points

$$x_i = i\Delta x, \quad v_j = j\Delta v - v_m, \quad \Delta x = \ell/I, \quad \Delta v = 2v_m/J .$$

We define the abbreviation

$$f_{ij}(t) \equiv f(x_i, v_j, t)$$

and the same for the Maxwellian f^M. For stability reasons, it is important to apply an implicit time scheme and (9.48) turns into

$$\frac{f_{ij}(t+\Delta t) - f_{ij}(t)}{\Delta t} + v_j \frac{f_{i+1,j}(t+\Delta t) - f_{i-1,j}(t+\Delta t)}{2\Delta x} = \frac{f_{ij}^M(t) - f_{ij}(t+\Delta t)}{\tau_0} \tag{9.50}$$

with the time step $\Delta t \ll \tau_0$. This can be written in the form

$$\sum_{k=0}^{I} M_{ik}^{(j)} f_{kj}(t+\Delta t) = \frac{f_{ij}(t)}{\Delta t} + \frac{f_{ij}^M(t)}{\tau_0} \qquad j = 0 \ldots J ,$$

where $M^{(j)}$ are tri-diagonal matrices

$$M_{ii}^{(j)} = \frac{1}{\Delta t} + \frac{1}{\tau_0}, \quad M_{i,i+1}^{(j)} = -M_{i,i-1}^{(j)} = \frac{v_j}{2\Delta x} ,$$

which can be inverted effectively by standard routines; for more details see [5].

Boundary conditions should be implemented in the distribution function and are dictated by the physics of the problem under consideration. If, for example, the velocity

u should vanish at say $x = 0$, f can be symmetrized at each time step according to

$$f(0, v, t + \Delta t) = f(0, -v, t + \Delta t) = \frac{1}{2}\left(f(0, v, t) + f(0, -v, t)\right).$$

9.4.3 Example: temperature step in an incompressible gas

Consider a one-dimensional gas having at a certain initial time t_0 a spatially step-like temperature shape

$$T_0(x) = 1 - a\tanh(b(x - x_0)) \tag{9.51}$$

with $a \ll 1$. Let us assume that density and pressure are homogeneous and stay so, i. e., the gas is incompressible and that the gas will perform no macroscopic motion during the whole process (valid for a solid or a liquid, but of course not for a gas). Thus we have

$$n = \text{const.}, \quad u = 0,$$

and we may put $n = 1$.

(i) Take at $t = 0$ the initial condition

$$f(t = 0) = f_0(x, v) = \frac{1}{\sqrt{2\pi T_0(x)}}\exp\left[-\frac{v^2}{2T_0(x)}\right].$$

(ii) Compute the temperature

$$T(x) = \langle v^2 \rangle .$$

At $t = 0$, this is the same as T_0.

(iii) Compute the Maxwellian

$$f^M = \frac{1}{\sqrt{2\pi T(x)}}\exp\left[-\frac{v^2}{2T(x)}\right].$$

At $t = 0$ this is the same as f_0.

(iv) Integrate numerically the 1D Boltzmann equation (9.48) from t to $t + \Delta t$.

(v) Go to (ii) and compute the new temperature with $f(\Delta t)$.

For the working fluid, we assume helium with $M = 4\,\text{g}$, for ℓ we take one meter and $T_0 = 300\,\text{K}$. This yields $\tau = 1.27 \cdot 10^{-3}$ s for the time scale. The grid has 200×200 mesh points; the time step is $\Delta t = 10^{-4}$. As initial condition, we take (9.51) with $a = 0.1$, $b = 25$, $x_0 = 1/2$. The relaxation time is chosen with $\tau_0 = 10^{-3}$, or $\tilde{\tau}_0 = 1.27 \cdot 10^{-6}$ s.

Figure 9.1, left frame, shows the results. The initial temperature step diffuses. Finally, an almost homogeneous temperature is reached that would asymptotically in time approach a constant value $T = 1$.

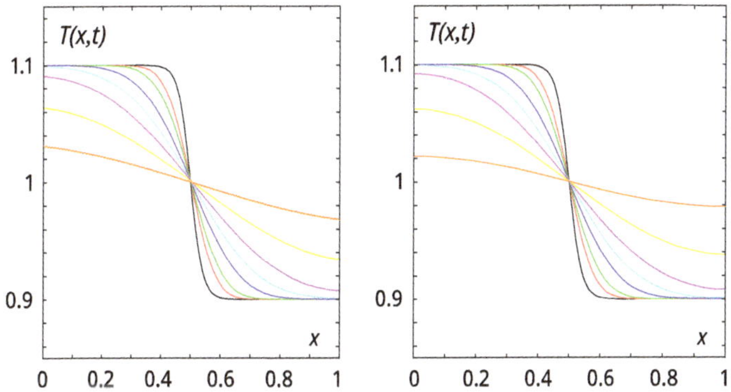

Figure 9.1: Left: Numerical integration of the 1D Boltzmann equation for an incompressible gas, $\tau_0 = 0.001$, at times $t = 0.01$ (black), 0.4 (red), 0.8 (green), 2 (blue), 4 (light blue), 8 (pink), 20 (yellow), 50 (brown). Right: Numerical solutions of the 1D heat equation for the same initial condition and the same times, $\kappa = 2.8\,\mathrm{m}^2/\mathrm{s}$.

The (unrealistic) long relaxation time leads, via the relation (9.36), to a rather low pressure of $p = \eta/\tau_0 \approx 16\,\mathrm{N/m}^2$ and finally to a large thermal diffusivity of $\kappa \approx 2.8\,\mathrm{m}^2/\mathrm{s}$. In the right frame of Figure 9.1, we show results found by numerically integrating the heat equation (4.1) in 1D. Indeed the curves look very similar.

In Figure 9.2, we show the difference

$$d(x, v) = f(x, v) - f^M(x, v) \tag{9.52}$$

at different consecutive times. In the beginning, the perturbations of the local equilibrium are large around the step but become more and more washed-out as time goes by. The integrated difference

$$\bar{d}(t) = \int_0^1 dx \int_{-v_m}^{v_m} dv\, |d(x, v)| \tag{9.53}$$

could serve as measure of the distance from thermal equilibrium (Figure 9.3).

9.4.4 Example: sound waves

In Section 9.2.3, we derived a wave equation for sound waves in a gas without viscosity and thermal conductivity. If both effects are included, equation (9.21) gets extra terms, which cause attenuation of the pressure wave. If 4th-order derivatives in space are ne-

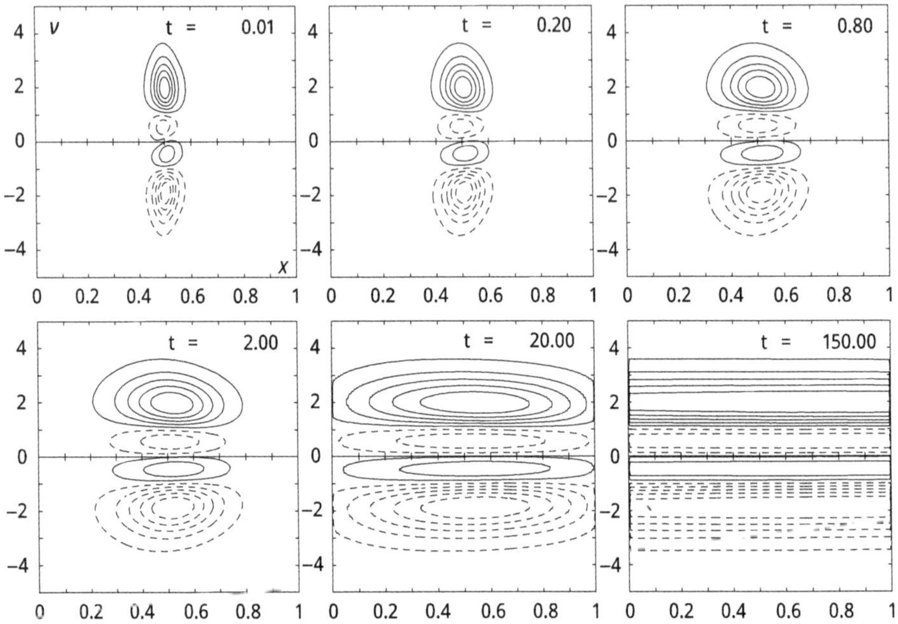

Figure 9.2: Contour lines of snapshots of d according to (9.52) in the xv-plane. Solid: $d > 0$, dashed: $d < 0$.

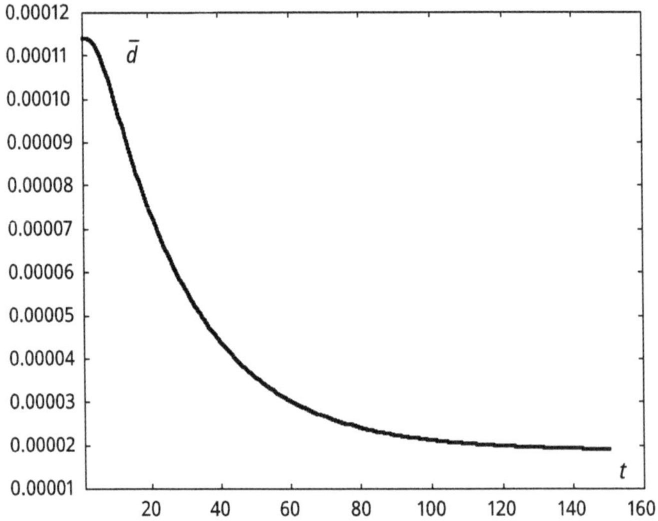

Figure 9.3: Distance to equilibrium $\bar{d}$ computed in (9.53) over time.

glected (long waves), the extended equation reads

$$\frac{\partial^2 p}{\partial t^2} - \frac{5}{3}\frac{p_0}{\rho_0}\,\nabla^2 p = (\nu + \kappa)\,\nabla^2 \frac{\partial p}{\partial t}\,. \tag{9.54}$$

Solving by

$$p(x,t) \sim e^{i(kx-\mu t)} \tag{9.55}$$

yields a damping rate of

$$\sigma = -\operatorname{Im}(\mu) = \frac{1}{2}(\nu + \kappa)\, k^2 \tag{9.56}$$

and the frequency

$$\omega = \operatorname{Re}(\mu) = \left(\frac{5}{3}\frac{p_0}{\rho_0}\right)^{1/2} k + O(k^3)\,, \tag{9.57}$$

leading to the same speed of sound $c_s = \omega/k$ as for the ideal case; see equation (9.22).

Now we shall try to use the Boltzmann equation to compute the propagation of a pressure pulse in a one-dimensional gas with BGK collision term. To account for pressure variations, the density becomes now a function of space and time and the macroscopic velocity u will not vanish anymore. Thus we have to take the full expression (9.47) for the local Maxwellian.

If we wish to study a single pressure pulse traveling, e. g., to the right side, we can take an initial condition, such as

$$n(x, t = 0) = 1 + a \exp(-x^2/b^2) \tag{9.58}$$

with a as an amplitude and b as the width of the distribution, simulating a compression close to the left wall. Furthermore, we assume $u(t = 0) = 0$ and $T(t = 0) = 1$ everywhere. The other parameters are $\ell = 1\,\mathrm{m}$, $T_0 = 300\,\mathrm{K}$, $\tau_0 = 10^{-4}$, $M = 4\,\mathrm{g}$. The numerical mesh size is 500×200 and the time step $\Delta t = 10^{-5}$.

Figure 9.4 shows the pressure distribution at several later times, Figure 9.5 the maximal value of p over time. The pressure pulse needs about 0.6 time units to travel from one wall to the other; the speed of sound found from (9.54) is $\sqrt{5/3}$ in the scaling we use here and agrees fairly well. The attenuation due to viscosity and thermal conductivity can be seen clearly. Comparing to an exponential decay (red line) expected from (9.55), the damping factor is about 0.55. Computing (9.56) in non-dimensional quantities and inserting (9.36) and (9.42) for $\nu = \eta/\rho_0$ and $\lambda = \kappa C R$, respectively, one finds

$$\sigma = \frac{4}{3}\,\tau_0\, k^2\,.$$

This results in the same value of 0.55 if we take $k = 63$, which would correspond to a wavelength of about 0.1, in good agreement with Figure 9.4.

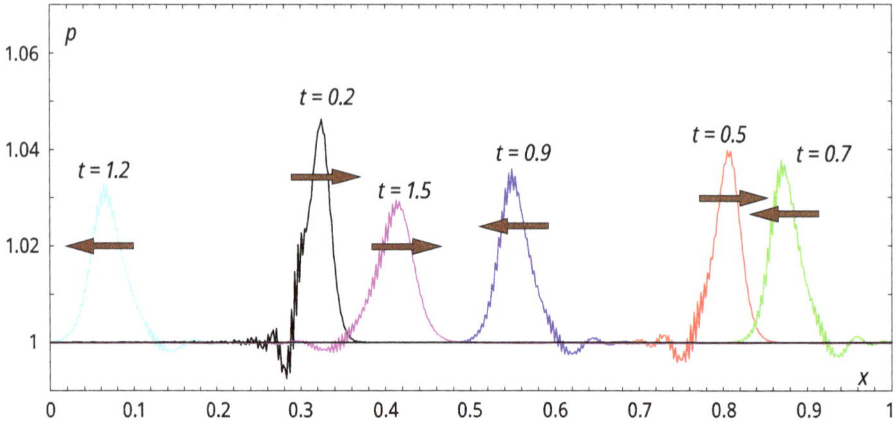

Figure 9.4: Propagation of the initial pressure wave (9.58) with $a = 0.1$, $b = 0.01$ in a one-dimensional gas. The waves at $t = 0.2,\ 0.5,\ 1.5$ (two reflections) travel to the right, the others to the left (one reflection).

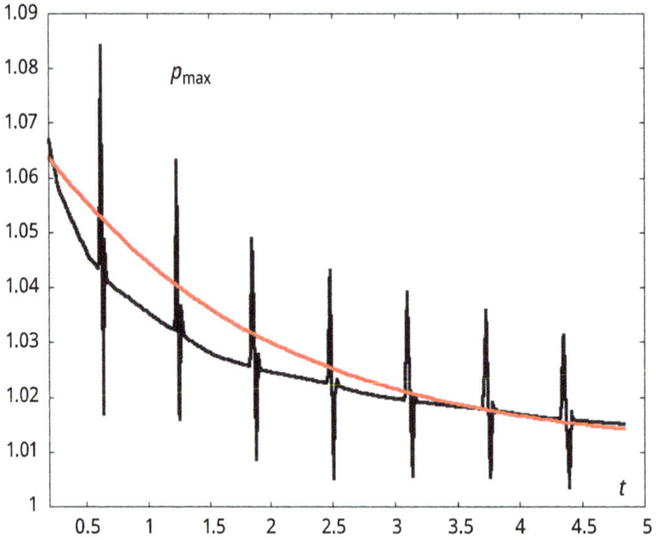

Figure 9.5: Black: Maximal values of the pressure over time. The peaks mark reflections at the walls. Red: Exponential decay with $\exp(-0.55\ t)$.

A Constants, material properties, and units

A.1 Constants

name	symbol	value	SI units
Avogadro's number	A_v, N_A	6.022×10^{23}	1/mol
Loschmidt's number	N_L	2.607×10^{25}	$1/m^3$ (at 0 °C, 1 atm)
Boltzmann constant	k, k_B	1.381×10^{-23}	J/K, $m^2 kg/(s^2 K)$
Planck constant	h	6.626×10^{-34}	J s, $m^2 kg/s$
reduced Planck const.	$\hbar$	1.105×10^{-34}	J s, $m^2 kg/s$
gas constant	R	8.3145	J/(K mol)
electron mass	m_e	9.11×10^{-31}	kg
proton mass	m_p	1.67×10^{-27}	kg
electron charge	e	1.602×10^{-19}	C
acceleration of gravity	g	9.81	m/s^2
speed of light (vacuum)	c	3.00×10^8	m/s

A.2 Material properties

A.2.1 Dry air, 30 °C

molar mass	28.9 g/mol
density	1.2 kg/m^3
specific heat c_v	718 J/(kg K)
specific heat c_p	1000 J/(kg K)
thermal conductivity	0.0261 J/(s m K)
thermal diffusivity	19×10^{-6} m^2/s
thermal expansion coef.	3.4×10^{-3} K^{-1}
kinematic viscosity	14×10^{-6} m^2/s
Prandtl number	0.74

A.2.2 Water

molar mass	18.02 g/mol
density	997 kg/m^3 (25 °C)
specific heat c_v	4140 J/(kg K)
specific heat c_p	4180 J/(kg K)
latent heat (boiling)	2.3×10^6 J/kg
thermal conductivity	0.56 J/(s m K)
thermal diffusivity	0.143×10^{-6} m^2/s
thermal expansion coef.	0.21×10^{-3} K^{-1}
kinematic viscosity	10^{-6} m^2/s
Prandtl number	7.0
surface tension (air)	0.07 N/m

https://doi.org/10.1515/9783110794137-012

A.2.3 Silicone oil 5 cSt

density	910 kg/m^3
specific heat c_p	1400 J/(kg K)
thermal conductivity	0.11 J/(s m K)
thermal diffusivity	0.07×10^{-6} m^2/s
thermal expansion coef.	1.1×10^{-3} K^{-1}
kinematic viscosity	5×10^{-6} m^2/s
Prandtl number	71.4
surface tension (air)	0.02 N/m

A.2.4 Ideal gas, monatomic

molar mass	M [kg]
density	$M N_L/N_A$ [kg/m^3]
specific heat c_v	$3R/(2M)$ [J/(kg K)]
specific heat c_p	$5R/(2M)$ [J/(kg K)]
thermal expansion coef.	$1/T$ [K^{-1}]
Prandtl number	1 (kinetic gas theory)

A.3 Other or older units

name	conversion
1 cal (calory)	4.19 J
1 atm (atmospheric pressure)	1.013×10^5 Pa
	= 1013 hPa
	= 760 torr
	= 760 mm Hg
	= 14.7 lbs/in^2
1 bar	10^5 Pa
1 cSt (centistokes)	10^{-6} m^2/s = 1 mm^2/s
1 eV (electron volt)	1.60×10^{-19} J
1 BTU (British thermal unit)	1055 J
1 hp (metric horsepower)	735.5 J/s

B Gaussian integrals

In kinetic gas theory, one often needs moments of Gauss functions of the form

$$J_n = \int_0^{\infty} dv\, v^n e^{-\frac{1}{2}m\beta v^2}, \quad n = 0, 1, 2, \dots . \tag{B.1}$$

Note that the limits are $0 \dots \infty$. Integrating over $-\infty \dots \infty$, the integrals for odd values of n vanish; those for even n have to be multiplied by two. For convenience, we list the first six J_n:

n	0	1	2	3	4	5
J_n	$\frac{(2\pi)^{1/2}}{2(m\beta)^{1/2}}$	$\frac{1}{m\beta}$	$\frac{(2\pi)^{1/2}}{2(m\beta)^{3/2}}$	$\frac{2}{(m\beta)^2}$	$\frac{3(2\pi)^{1/2}}{2(m\beta)^{5/2}}$	$\frac{8}{(m\beta)^3}$

(B.2)

Using the table, we can, for instance, compute the 3D integral:

$$\begin{aligned}
&\int_{-\infty}^{\infty} dv_x \int_{-\infty}^{\infty} dv_y \int_0^{\infty} dv_z v^2 v_z e^{-\frac{1}{2}m\beta v^2} \\
&= \int_{-\infty}^{\infty} dv_x v_x^2 e^{-\frac{1}{2}m\beta v_x^2} \int_{-\infty}^{\infty} dv_y e^{-\frac{1}{2}m\beta v_y^2} \int_0^{\infty} dv_z v_z e^{-\frac{1}{2}m\beta v_z^2} \\
&\quad + \int_{-\infty}^{\infty} dv_x e^{-\frac{1}{2}m\beta v_x^2} \int_{-\infty}^{\infty} dv_y v_y^2 e^{-\frac{1}{2}m\beta v_y^2} \int_0^{\infty} dv_z v_z e^{-\frac{1}{2}m\beta v_z^2} \\
&\quad + \int_{-\infty}^{\infty} dv_x e^{-\frac{1}{2}m\beta v_x^2} \int_{-\infty}^{\infty} dv_y e^{-\frac{1}{2}m\beta v_y^2} \int_0^{\infty} dv_z v_z^3 e^{-\frac{1}{2}m\beta v_z^2} \\
&= 4J_0J_1J_2 + 4J_0J_1J_2 + 4J_0^2J_3 = \frac{8\pi}{(m\beta)^3} .
\end{aligned} \tag{B.3}$$

The norm reads

$$\mathcal{N} = \int_{-\infty}^{\infty} dv_x \int_{-\infty}^{\infty} dv_y \int_{-\infty}^{\infty} dv_z\, e^{-\frac{1}{2}m\beta v^2} = 4\pi \int_0^{\infty} dv\, v^2 e^{-\frac{1}{2}m\beta v^2} = 4\pi J_2 = \left(\frac{2\pi}{m\beta}\right)^{3/2} . \tag{B.4}$$

https://doi.org/10.1515/9783110794137-013

C Theorems and relations from vector calculus

We compile some important relations used throughout the book. The notion "tensor" is, if not stated otherwise, used for a tensor of the second order.

C.1 Definitions and products

The relations in components are valid for cartesian coordinates.

C.1.1 Definitions

1. The Kronecker symbol or unit tensor is

$$(1)_{ij} = \delta_{ij} = \begin{cases} 1 & \text{for} \quad i = j, \\ 0 & \text{else.} \end{cases} \tag{C.1}$$

2. The completely antisymmetric tensor of the third order, or Levi-Civita tensor is

$$\epsilon_{ijk} = \begin{cases} 1 & \text{if } i, j, k \text{ even permutation of } 1, 2, 3, \\ -1 & \text{if } i, j, k \text{ odd permutation of } 1, 2, 3, \\ 0 & \text{if two or three indices are equal.} \end{cases} \tag{C.2}$$

C.1.2 Products

1. Vectors:
   scalar product

$$\vec{a} \cdot \vec{b} = \sum_i a_i b_i; \tag{C.3}$$

   vector product, cross product

$$(\vec{a} \times \vec{b})_k = \sum_{ij} \epsilon_{kij} a_i b_j; \tag{C.4}$$

   dyadic product

$$(\vec{a} \otimes \vec{b})_{ij} = a_i b_j. \tag{C.5}$$

https://doi.org/10.1515/9783110794137-014

2. Tensors:
   inner (double) product

$$\underline{A} : \underline{B} = \sum_{ij} A_{ij} B_{ij}; \tag{C.6}$$

   tensor multiplication

$$(\underline{A} \cdot \underline{B})_{ij} = \sum_{k} A_{ik} B_{kj}; \tag{C.7}$$

3. Higher products:
   scalar triple product

$$\begin{aligned} \vec{a} \cdot (\vec{b} \times \vec{c}) &= \vec{b} \cdot (\vec{c} \times \vec{a}) = \vec{c} \cdot (\vec{a} \times \vec{b}); \\ \sum_{ijk} \epsilon_{ijk} a_i b_j c_k &= \sum_{ijk} \epsilon_{jki} a_i b_j c_k = \sum_{ijk} \epsilon_{kij} a_i b_j c_k; \end{aligned} \tag{C.8}$$

   scalar product of vector products

$$\begin{aligned} (\vec{a} \times \vec{b}) \cdot (\vec{c} \times \vec{d}) &= (\vec{a} \cdot \vec{c})\,(\vec{b} \cdot \vec{d}) - (\vec{a} \cdot \vec{d})\,(\vec{b} \cdot \vec{c}); \\ \sum_{ijk\ell m} \epsilon_{ijk} \epsilon_{i\ell m} a_j b_k c_\ell d_m &= \sum_{ij} a_i c_i b_j d_j - \sum_{ij} a_i d_i b_j c_j; \end{aligned} \tag{C.9}$$

   double vector product

$$\begin{aligned} \vec{a} \times (\vec{b} \times \vec{c}) &= (\vec{a} \cdot \vec{c})\,\vec{b} - (\vec{a} \cdot \vec{b})\,\vec{c}; \\ \sum_{jk\ell m} \epsilon_{ijk} \epsilon_{k\ell m} a_j b_\ell c_m &= \sum_{j} a_j c_j b_i - \sum_{j} a_j b_j c_i. \end{aligned} \tag{C.10}$$

C.2 Differentiation and integration of vector fields

Next we allow for spacially varying vector and tensor components. Let $\Psi(\vec{r})$ be a differentiable scalar field and $\vec{v}(\vec{r})$, $\vec{w}(\vec{r})$ vector fields. Again, the relations in components are valid for Cartesian coordinates.

C.2.1 Relations

1. For arbitrary Ψ,

$$\begin{aligned} \nabla \times (\nabla \Psi) &= 0, \\ \sum_{jk} \epsilon_{ijk} \partial_j \partial_k \Psi &= 0, \end{aligned} \tag{C.11}$$

and

$$\nabla \cdot (\nabla \times \vec{v}) = 0, \tag{C.12}$$
$$\sum_{ijk} \epsilon_{ijk} \partial_i \partial_j v_k = 0.$$

2. Transport term:

$$(\vec{v} \cdot \nabla)\, \vec{v} = \vec{v} \cdot (\nabla \otimes \vec{v}) = \frac{1}{2} \nabla v^2 - \vec{v} \times (\nabla \times \vec{v}), \tag{C.13}$$
$$\sum_j v_j \partial_j v_i = \frac{1}{2} \partial_i \sum_j v_j^2 - \epsilon_{ijk} \epsilon_{k\ell m} v_j \partial_\ell v_m;$$
$$\left((\vec{v} \cdot \nabla)\, \vec{v} \right)_k = \nabla \cdot (v_k \vec{v}) - v_k \nabla \cdot \vec{v}, \tag{C.14}$$
$$\sum_j v_j \partial_j v_i = \sum_i \partial_i v_k v_i - v_k \sum_i \partial_i v_i.$$

For incompressible flow $\nabla \cdot \vec{v} = 0$,

$$\left((\vec{v} \cdot \nabla)\, \vec{v} \right)_k = \nabla \cdot (v_k \vec{v}). \tag{C.15}$$

3. Double curl:

$$\nabla \times (\nabla \times \vec{v}) = \nabla (\nabla \cdot \vec{v}) - \nabla^2 \vec{v}, \tag{C.16}$$
$$\sum_{jk\ell m} \epsilon_{ijk} \epsilon_{k\ell m} \partial_j \partial_\ell v_m = \partial_i \sum_j \partial_j v_j - \sum_j \partial_j^2 v_i.$$

4. Scalar triple product:

$$\nabla \cdot (\vec{v} \times \vec{w}) = \vec{w} \cdot (\nabla \times \vec{v}) - \vec{v} \cdot (\nabla \times \vec{w}), \tag{C.17}$$
$$\sum_{ijk} \epsilon_{ijk} \partial_i (v_j w_k) = \sum_{ijk} \epsilon_{ijk} w_i \partial_j v_k - \sum_{ijk} \epsilon_{ijk} v_i \partial_j w_k.$$

5. Gradient of a scalar product:

$$\nabla \cdot (\vec{v} \cdot \vec{w}) = (\vec{v} \cdot \nabla) \vec{w} + (\vec{w} \cdot \nabla) \vec{v} + \vec{v} \times (\nabla \times \vec{w}) + \vec{w} \times (\nabla \times \vec{v}),$$
$$\partial_i \sum_j (v_j w_j) = \sum_j v_j \partial_j w_i + \sum_j w_j \partial_j v_i + \sum_{jk\ell m} \epsilon_{ijk} \epsilon_{k\ell m} v_i \partial_\ell w_m$$
$$+ \sum_{jk\ell m} \epsilon_{ijk} \epsilon_{k\ell m} w_i \partial_\ell v_m. \tag{C.18}$$

6. Double vector product:

$$\nabla \times (\vec{v} \times \vec{w}) = \vec{v}\, \nabla \times \vec{w} - \vec{w}\, \nabla \times \vec{v} + (\vec{w} \cdot \nabla)\, \vec{v} - (\vec{v} \cdot \nabla)\, \vec{w},$$
$$\sum_{jk\ell m} \epsilon_{ijk} \epsilon_{k\ell m} \partial_j (v_\ell w_m) = v_i \sum_j \partial_j w_j - w_i \sum_j \partial_j v_j$$
$$+ \sum_j w_j \partial_j v_i - \sum_j v_j \partial_j w_i. \tag{C.19}$$

7. Differentiation of products of scalar and vector fields:

$$\nabla \cdot (\Psi \vec{v}) = \vec{v} \cdot \nabla \Psi + \Psi \nabla \cdot \vec{v}, \tag{C.20}$$
$$\sum_i \partial_i (\Psi v_i) = \sum_i v_i \partial_i \Psi + \Psi \sum_i \partial_i v_i;$$
$$\nabla \times (\Psi \vec{v}) = (\nabla \Psi) \times \vec{v} + \Psi (\nabla \times \vec{v}), \tag{C.21}$$
$$\sum_{jk} \epsilon_{ijk} \partial_j (\Psi v_k) = \sum_{jk} \epsilon_{ijk} (\partial_j \Psi) v_k + \sum_{jk} \epsilon_{ijk} \Psi \partial_j v_k .$$

C.3 Integral theorems

Let $\vec{v}$ be a differentiable vector field, and let Ψ and Φ differentiable scalar functions of $\vec{r}$. The 3D volume element (scalar) is denoted with $d^3\vec{r}$, the 2D surface element (vector) with $d^2\vec{f}$, and the 1D line element (vector) with $d\vec{s}$.

C.3.1 Volume and surface integrals

Let V be an arbitrary, simply connected volume, and let $F(V)$ be a closed surface of V. The Gauss theorem reads

$$\int_V \nabla \cdot \vec{v}\, d^3\vec{r} = \oint_{F(V)} \vec{v}\, d^2\vec{f} . \tag{C.22}$$

Two other, related theorems are

$$\int_V \nabla \Psi\, d^3\vec{r} = \oint_{F(V)} \Psi\, d^2\vec{f} \tag{C.23}$$

and

$$\int_V (\nabla \times \vec{v})\, d^3\vec{r} = - \oint_{F(V)} \vec{v} \times d^2\vec{f} . \tag{C.24}$$

C.3.2 Surface and line integrals

Let $C(F)$ be a closed line confining the simply connected area F. Stokes' theorem reads

$$\int_F (\nabla \times \vec{v})\, d^2\vec{f} = \oint_{C(F)} \vec{v}\, d\vec{s} = \Gamma \tag{C.25}$$

with Γ defined as the circulation of $\vec{v}$. A related theorem is

$$\int_F (\nabla\Psi) \times d^2\vec{f} = -\oint_{C(F)} \Psi \, d\vec{s} \,. \tag{C.26}$$

C.4 Other coordinate systems

We list the important differential operators for Cartesian, cylindrical, and spherical coordinate systems and their effect on scalar, vector, and tensor fields.

C.4.1 Relations

C.4.1.1 Cylindrical coordinates

$$x = r\cos\varphi, \quad y = r\sin\varphi\,, \tag{C.27}$$

$$\hat{e}_r = \cos\varphi\,\hat{e}_x + \sin\varphi\,\hat{e}_y,$$
$$\hat{e}_\varphi = -r\sin\varphi\,\hat{e}_x + r\cos\varphi\,\hat{e}_y\,. \tag{C.28}$$

Vectors:

$$\vec{v} = v_x\hat{e}_x + v_y\hat{e}_y + v_z\hat{e}_z = v_r\hat{e}_r + v_\varphi\hat{e}_\varphi + v_z\hat{e}_z \tag{C.29}$$

with

$$v_x = \vec{v}\cdot\hat{e}_x, \quad v_y = \vec{v}\cdot\hat{e}_y, \quad v_z = \vec{v}\cdot\hat{e}_z$$

and

$$v_r = \vec{v}\cdot\hat{e}_r, \quad v_\varphi = \frac{1}{r^2}\vec{v}\cdot\hat{e}_\varphi, \quad v_z = \vec{v}\cdot\hat{e}_z\,.$$

For illustration, see Figure C.1, left frame.

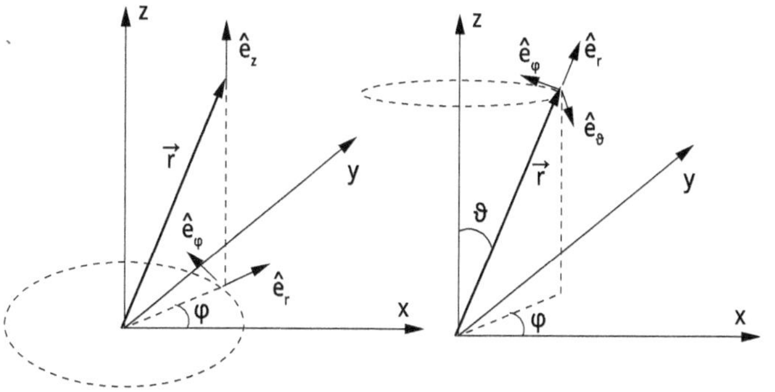

Figure C.1: Unit vectors and coordinates for cylindrical (left) and spherical (right) systems.

C.4.1.2 Spherical coordinates

$$x = r\cos\varphi\sin\vartheta, \quad y = r\sin\varphi\sin\vartheta, \quad z = r\cos\vartheta\,, \tag{C.30}$$

$$\begin{aligned} \hat{e}_r &= \cos\varphi\sin\vartheta\,\hat{e}_x + \sin\varphi\sin\vartheta\,\hat{e}_y + \cos\vartheta\,\hat{e}_z, \\ \hat{e}_\varphi &= -r\sin\varphi\sin\vartheta\,\hat{e}_x + r\cos\varphi\sin\vartheta\,\hat{e}_y, \\ \hat{e}_\vartheta &= r\cos\varphi\cos\vartheta\,\hat{e}_x + r\sin\varphi\cos\vartheta\,\hat{e}_y - r\sin\vartheta\,\hat{e}_z\,. \end{aligned} \tag{C.31}$$

Vectors:

$$\vec{v} = v_r\hat{e}_r + v_\varphi\hat{e}_\varphi + v_\vartheta\hat{e}_\vartheta \tag{C.32}$$

with

$$v_r = \vec{v}\cdot\hat{e}_r, \quad v_\varphi = \frac{1}{r^2}\vec{v}\cdot\hat{e}_\varphi, \quad v_\vartheta = \frac{1}{r^2}\vec{v}\cdot\hat{e}_\vartheta.$$

For illustration, see Figure C.1, right frame.

The same representation is valid for tensors:

$$\underline{T} = \sum_{ij} T_{ij}\hat{e}_i \otimes \hat{e}_j\,, \tag{C.33}$$

where i, j stands for x, y, z, but also for r, φ, z or for r, φ, and ϑ.

C.4.2 Gradient

C.4.2.1 Cartesian coordinates

$$\nabla = \hat{e}_x\partial_x + \hat{e}_y\partial_y + \hat{e}_z\partial_z, \tag{C.34}$$

$$\nabla U = \nabla U = \partial_x U\hat{e}_x + \partial_y U\hat{e}_y + \partial_z U\hat{e}_z, \tag{C.35}$$

$$\nabla\otimes\vec{v} = \begin{pmatrix} \partial_x v_x & \partial_x v_y & \partial_x v_z \\ \partial_y v_x & \partial_y v_y & \partial_y v_z \\ \partial_z v_x & \partial_z v_y & \partial_z v_z \end{pmatrix}. \tag{C.36}$$

C.4.2.2 Cylindrical coordinates

$$\begin{aligned} \nabla &= \hat{e}_r\partial_r + \hat{e}_\varphi\frac{1}{r}\partial_\varphi + \hat{e}_z\partial_z, \\ \nabla U &= \nabla U = (\partial_r U)\hat{e}_r + \frac{1}{r}(\partial_\varphi U)\hat{e}_\varphi + (\partial_z U)\hat{e}_z\,. \end{aligned} \tag{C.37}$$

$$\nabla \otimes \vec{v} = \begin{pmatrix} \partial_r v_r & \partial_r v_\varphi & \partial_r v_z \\ \frac{1}{r}(\partial_\varphi v_r - v_\varphi) & \frac{1}{r}(\partial_\varphi v_\varphi + v_r) & \frac{1}{r}\partial_\varphi v_z \\ \partial_z v_r & \partial_z v_\varphi & \partial_z v_z \end{pmatrix}. \tag{C.38}$$

C.4.2.3 Spherical coordinates

$$\nabla = \hat{e}_r \partial_r + \hat{e}_\varphi \frac{1}{r \sin\vartheta}\partial_\varphi + \hat{e}_\vartheta \frac{1}{r}\partial_\vartheta, \tag{C.39}$$

$$\nabla U = \nabla U = (\partial_r U)\hat{e}_r + \frac{1}{r\sin\vartheta}(\partial_\varphi U)\hat{e}_\varphi + \frac{1}{r}(\partial_\vartheta U)\hat{e}_\vartheta \; ;$$

$$\nabla \otimes \vec{v} = \begin{pmatrix} \partial_r v_r & \partial_r v_\varphi & \partial_r v_\vartheta \\ \frac{1}{r}\left(\frac{\partial_\varphi v_r}{\sin\vartheta} - v_\varphi\right) & \frac{1}{r\sin\vartheta}\partial_\varphi v_\varphi + \frac{v_r}{r} + \frac{v_\vartheta \cot\vartheta}{r} & \frac{1}{r}\left(\frac{\partial_\varphi v_\vartheta}{\sin\vartheta} - v_\varphi \cot\vartheta\right) \\ \frac{1}{r}(\partial_\vartheta v_r - v_\vartheta) & \frac{1}{r}\partial_\vartheta v_\varphi & \frac{1}{r}(\partial_\vartheta v_\vartheta + v_r) \end{pmatrix}. \tag{C.40}$$

C.4.3 Divergence

C.4.3.1 Cartesian coordinates

$$\nabla \cdot \vec{v} = \partial_x v_x + \partial_y v_y + \partial_z v_z, \tag{C.41}$$

$$\begin{aligned} \nabla \cdot \underline{T} = {} & [\partial_x t_{xx} + \partial_y t_{yx} + \partial_z t_{zx}]\,\hat{e}_x \\ & + [\partial_x t_{xy} + \partial_y t_{yy} + \partial_z t_{zy}]\,\hat{e}_y \\ & + [\partial_x t_{xz} + \partial_y t_{yz} + \partial_z t_{zz}]\,\hat{e}_z. \end{aligned} \tag{C.42}$$

C.4.3.2 Cylindrical coordinates

$$\begin{aligned} \nabla \cdot \vec{v} &= \partial_r v_r + \frac{1}{r}(\partial_\varphi v_\varphi + v_r) + \partial_z v_z \\ &= \frac{1}{r}\partial_r(r v_r) + \frac{1}{r}\partial_\varphi v_\varphi + \partial_z v_z, \end{aligned} \tag{C.43}$$

$$\nabla \cdot \underline{T} = \left[\partial_r t_{rr} + \frac{1}{r}(\partial_\varphi t_{\varphi r} + t_{rr} - t_{\varphi\varphi}) + \partial_z t_{zr}\right]\hat{e}_r$$

$$+ \left[\partial_r t_{r\varphi} + \frac{1}{r}(\partial_\varphi t_{\varphi\varphi} + t_{r\varphi} + t_{\varphi r}) + \partial_z t_{z\varphi}\right]\hat{e}_\varphi \tag{C.44}$$

$$+ \left[\partial_r t_{rz} + \frac{1}{r}(\partial_\varphi t_{\varphi z} + t_{rz}) + \partial_z t_{zz}\right]\hat{e}_z. \tag{C.45}$$

C.4.3.3 Spherical coordinates

$$\begin{aligned}\nabla\cdot\vec{v} &= \partial_r v_r + \frac{1}{r\sin\vartheta}\partial_\varphi v_\varphi + \frac{1}{r}\partial_\vartheta v_\vartheta + 2\frac{v_r}{r} + \frac{v_\vartheta\cot\vartheta}{r}\\ &= \frac{1}{r^2}\partial_r(r^2 v_r) + \frac{1}{r\sin\vartheta}\partial_\varphi v_\varphi + \frac{1}{r\sin\vartheta}\partial_\vartheta(v_\vartheta\sin\vartheta);\end{aligned} \tag{C.46}$$

$$\begin{aligned}\nabla\cdot\underline{T} = &\left[\frac{1}{r^2}\partial_r(r^2 t_{rr}) + \frac{1}{r\sin\vartheta}\partial_\varphi t_{\varphi r} + \frac{1}{r\sin\vartheta}\partial_\vartheta(t_{\vartheta r}\sin\vartheta) - \frac{1}{r}(t_{\vartheta\vartheta}+t_{\varphi\varphi})\right]\hat{e}_r\\ &+\left[\frac{1}{r^3}\partial_r(r^3 t_{r\varphi}) + \frac{1}{r\sin\vartheta}\partial_\varphi t_{\varphi\varphi} + \frac{1}{r\sin\vartheta}\partial_\vartheta(t_{\vartheta\varphi}\sin\vartheta) + \frac{1}{r}(t_{\varphi r} - t_{r\varphi} + t_{\varphi\vartheta}\cot\vartheta)\right]\hat{e}_\varphi\\ &+\left[\frac{1}{r^3}\partial_r(r^3 t_{r\vartheta}) + \frac{1}{r\sin\vartheta}\partial_\varphi t_{\varphi\vartheta} + \frac{1}{r\sin\vartheta}\partial_\vartheta(t_{\vartheta\vartheta}\sin\vartheta) + \frac{1}{r}(t_{\vartheta r} - t_{r\vartheta} - t_{\varphi\varphi}\cot\vartheta)\right]\hat{e}_\vartheta.\end{aligned} \tag{C.47}$$

C.4.4 Curl

C.4.4.1 Cartesian coordinates

$$\nabla\times\vec{v} = (\partial_y v_z - \partial_z v_y)\,\hat{e}_x + (\partial_z v_x - \partial_x v_z)\,\hat{e}_y + (\partial_x v_y - \partial_y v_x)\,\hat{e}_z. \tag{C.48}$$

C.4.4.2 Cylindrical coordinates

$$\nabla\times\vec{v} = \left(\frac{1}{r}\partial_\varphi v_z - \partial_z v_\varphi\right)\hat{e}_r + (\partial_z v_r - \partial_r v_z)\,\hat{e}_\varphi + \frac{1}{r}(\partial_r(r v_\varphi) - \partial_\varphi v_r)\,\hat{e}_z. \tag{C.49}$$

C.4.4.3 Spherical coordinates

$$\begin{aligned}\nabla\cdot\vec{v} = &\frac{1}{r\sin\vartheta}[\partial_\vartheta(v_\varphi\sin\vartheta) - \partial_\varphi v_\vartheta]\,\hat{e}_r\\ &+\frac{1}{r}[\partial_r(r v_\vartheta) - \partial_\vartheta v_r]\,\hat{e}_\varphi\\ &+\frac{1}{r}\left[\frac{1}{\sin\vartheta}\partial_\varphi v_r - \partial_r(r v_\varphi)\right]\hat{e}_\vartheta.\end{aligned} \tag{C.50}$$

C.4.5 Laplacian

C.4.5.1 Cartesian coordinates

$$\nabla^2 U = \partial^2_{xx}U + \partial^2_{yy}U + \partial^2_{zz}U, \tag{C.51}$$

$$\nabla^2\vec{v} = \nabla^2 v_x\,\hat{e}_x + \nabla^2 v_y\,\hat{e}_y + \nabla^2 v_z\,\hat{e}_z. \tag{C.52}$$

C.4.5.2 Cylindrical coordinates

$$\nabla^2 U = \frac{1}{r}\partial_r(r\partial_r U) + \frac{1}{r^2}\partial^2_{\varphi\varphi}U + \partial^2_{zz}U = \partial^2_{rr}U + \frac{1}{r}\partial_r U + \frac{1}{r^2}\partial^2_{\varphi\varphi}U + \partial^2_{zz}U. \tag{C.53}$$

$$\nabla^2\vec{v} = \left[\nabla^2 v_r - \frac{1}{r^2}v_r - \frac{2}{r^2}\partial_\varphi v_\varphi\right]\hat{e}_r + \left[\nabla^2 v_\varphi - \frac{1}{r^2}v_\varphi + \frac{2}{r^2}\partial_\varphi v_r\right]\hat{e}_\varphi + \nabla^2 v_z\,\hat{e}_z. \tag{C.54}$$

C.4.5.3 Spherical coordinates

$$\begin{aligned}\nabla^2 U &= \frac{1}{r^2}\partial_r(r^2\partial_r U) + \frac{1}{r^2\sin^2\vartheta}\partial^2_{\varphi\varphi}U + \frac{1}{r^2\sin\vartheta}\partial_\vartheta(\sin\vartheta\,\partial_\vartheta U)\\ &= \partial^2_{rr}U + \frac{2}{r}\partial_r U + \frac{1}{r^2\sin^2\vartheta}\partial^2_{\varphi\varphi}U + \frac{1}{r^2}\partial^2_{\vartheta\vartheta}U + \frac{\cot\vartheta}{r^2}\partial_\vartheta U;\end{aligned} \tag{C.55}$$

$$\begin{aligned}\nabla^2\vec{v} &= \left[\nabla^2 v_r - \frac{2}{r^2}v_r - \frac{2}{r^2\sin\vartheta}\partial_\varphi v_\varphi - \frac{2}{r^2}\partial_\vartheta v_\vartheta - \frac{2}{r^2}v_\vartheta\cot\vartheta\right]\hat{e}_r\\ &+ \left[\nabla^2 v_\varphi - \frac{v_\varphi}{r^2\sin^2\vartheta} + \frac{2}{r^2\sin^2\vartheta}\partial_\varphi v_r + \frac{2\cos\vartheta}{r^2\sin^2\vartheta}\partial_\varphi v_\vartheta\right]\hat{e}_\varphi\\ &+ \left[\nabla^2 v_\vartheta + \frac{2}{r^2}\partial_\vartheta v_r - \frac{2}{r^2\sin\vartheta}\partial_\varphi v_\varphi - \frac{v_\vartheta}{r^2\sin^2\vartheta}\right]\hat{e}_\vartheta.\end{aligned} \tag{C.56}$$

C.5 Hydrodynamic basic equations

Wie consider only the case of incompressible liquids

C.5.1 Navier–Stokes equations

C.5.1.1 Cylindrical coordinates

$$\begin{aligned}\rho\left[\frac{dv_r}{dt} - \frac{v_\varphi^2}{r}\right] &= -\partial_r p + f_r + \eta\left[\nabla^2 v_r - \frac{v_r}{r^2} - \frac{2}{r^2}\partial_\varphi v_\varphi\right];\\ \rho\left[\frac{dv_\varphi}{dt} + \frac{v_r v_\varphi}{r}\right] &= -\frac{1}{r}\partial_\varphi p + f_\varphi + \eta\left[\nabla^2 v_\varphi - \frac{v_\varphi}{r^2} + \frac{2}{r^2}\partial_\varphi v_r\right];\\ \rho\frac{dv_z}{dt} &= -\partial_z p + f_z + \eta\nabla^2 v_z\end{aligned} \tag{C.57}$$

with the material derivative

$$\frac{d}{dt} = \partial_t + v_r\partial_r + \frac{1}{r}v_\varphi\partial_\varphi + v_z\partial_z. \tag{C.58}$$

C.5.1.2 Spherical coordinates

$$\rho\left[\frac{dv_r}{dt} - \frac{v_\varphi^2 + v_\vartheta^2}{r}\right] = -\partial_r p + f_r + \eta\left[\nabla^2 v_r - \frac{2}{r^2}\left(v_r + \partial_\vartheta v_\vartheta + v_\theta \cot\vartheta + \frac{1}{\sin\vartheta}\partial_\varphi v_\varphi\right)\right]; \tag{C.59}$$

$$\rho\left[\frac{dv_\varphi}{dt} + \frac{v_r v_\varphi + v_\vartheta v_\varphi \cot\vartheta}{r}\right] = -\frac{1}{r\sin\vartheta}\partial_\varphi p + f_\varphi + \eta\left[\nabla^2 v_\varphi + \frac{1}{r^2\sin^2\vartheta}(-v_\varphi + \partial_\varphi v_r + 2\cos\vartheta\partial_\varphi v_\vartheta)\right];$$

$$\rho\left[\frac{dv_\vartheta}{dt} + \frac{v_r v_\vartheta - v_\varphi^2\cot\vartheta}{r}\right] = -\frac{1}{r}\partial_\vartheta p + f_\vartheta + \eta\left[\nabla^2 v_\vartheta + \frac{2}{r^2}\left(\partial_\vartheta v_r - \frac{v_\vartheta}{\sin^2\vartheta} - \frac{2\cos\vartheta}{r^2\sin^2\vartheta}\partial_\varphi v_\varphi\right)\right]$$

with the material derivative

$$\frac{d}{dt} = \partial_t + v_r\partial_r + \frac{v_\varphi}{r\sin\vartheta}\partial_\varphi + \frac{1}{r}v_\vartheta\partial_\vartheta. \tag{C.60}$$

C.5.2 Incompressibility condition

For incompressible fluids, the continuity equation reduces to $\nabla\cdot\vec{v} = 0$.

C.5.2.1 Cylindrical coordinates

$$\frac{1}{r}\partial_r(rv_r) + \frac{1}{r}\partial_\varphi v_\varphi + \partial_z v_z = 0. \tag{C.61}$$

C.5.2.2 Spherical coordinates

$$\frac{1}{r^2}\partial_r(r^2 v_r) + \frac{1}{r\sin\vartheta}\partial_\varphi v_\varphi + \frac{1}{r\sin\vartheta}\partial_\vartheta(\sin\vartheta\; v_\vartheta) = 0. \tag{C.62}$$

Bibliography

[1] H. B. Callen, *Thermodynamics and an Introduction to Thermostatistics*, Wiley (1985).
[2] D. S. Lemons, *A Student's Guide to Entropy*, Cambridge (2013).
[3] S. R. de Groot, P.Mazur, *Non-Equilibrium Thermodynamics*, Dover (1984).
[4] L. Onsager, *Reciprocal relations in irreversible processes. I*, Phys. Rev. 37, 405 (1931).
[5] M. Bestehorn, *Computational Physics*, De Gruyter, 2nd edn (2023).
[6] H. Haken, *Synergetics: Introduction and Advanced Topics*, Springer (2012).
[7] A. M. Turing: *The chemical basis of morphogenesis*, Phil. Trans. R. Soc. London B 237, 37 (1952).
[8] L. D. Landau, E. M. Lifshitz, *Fluid Mechanics, Vol. 6*, Butterworth-Heinemann, 2 edn (1987).
[9] C. A. J. Fletcher, *Computational Techniques for Fluid Dynamics 1, 2*, Springer (2013).
[10] C.-H. Bruneau, M. Saad, *The 2D lid-driven cavity problem revisited*, Computers & Fluids 35, 326 (2006).
[11] P. Kolodner, A. Passner, H. L. Williams, C. M. Surko, *The transition to finite-amplitude traveling-wave convection in binary fluid mixtures*, Nuclear Phys. (Proc. Sup.) B 2, 97 (1987).
[12] J. P. Peixoto, A. H. Oort, *Physics of Climate*, American Institute of Physics (1995).
[13] T. I. Gombosi, *Gaskinetic Theory*, Cambridge Univ. Press (1994).
[14] R. P. Feynman, R. B. Leighton, M. Sands, *The Feynman Lectures on Physics*, Basic Books, New Millenium edn (2011).
[15] L. D. Landau, E. M. Lifshitz, *Mechanics, Vol. 1*, Butterworth-Heinemann, 3 edn (1976).

https://doi.org/10.1515/9783110794137-015

Index

https://doi.org/10.1515/9783110794137-016

www.ingramcontent.com/pod-product-compliance
Lightning Source LLC
Chambersburg PA
CBHW061406210326
41598CB00035B/6120
9783110794045